Brief Contents

Table of Problem-Solving Strategies

Note for users of the five-volume edition:
Volume 1 (pp. 1–443) includes chapters 1–15.
Volume 2 (pp. 444–559) includes chapters 16–19.
Volume 3 (pp. 560–719) includes chapters 20–24.
Volume 4 (pp. 720–1101) includes chapters 25–36.
Volume 5 (pp. 1102–1279) includes chapters 36–42.

Chapters 37–42 are not in the Standard Edition.

THIRD EDITION

physics

FOR SCIENTISTS AND ENGINEERS
a strategic approach

VOLUME 3

randall d. knight

California Polytechnic State University
San Luis Obispo

Boston Columbus Indianapolis New York San Francisco Upper Saddle River
Amsterdam Cape Town Dubai London Madrid Milan Munich Paris Montreal Toronto
Delhi Mexico City Sao Paulo Sydney Hong Kong Seoul Singapore Taipei Tokyo

Publisher:	James Smith
Senior Development Editor:	Alice Houston, Ph.D.
Senior Project Editor:	Martha Steele
Assistant Editor:	Peter Alston
Media Producer:	Kelly Reed
Senior Administrative Assistant:	Cathy Glenn
Director of Marketing:	Christy Lesko
Executive Marketing Manager:	Kerry McGinnis
Managing Editor:	Corinne Benson
Production Project Manager:	Beth Collins
Production Management, Composition, and Interior Design:	Cenveo Publisher Services/Nesbitt Graphics, Inc.
Illustrations:	Rolin Graphics
Cover Design:	Yvo Riezebos Design
Manufacturing Buyer:	Jeff Sargent
Photo Research:	Eric Schrader
Image Lead:	Maya Melenchuk
Cover Printer:	Lehigh-Phoenix
Text Printer and Binder:	R.R. Donnelley/Willard
Cover Image:	Composite illustration by Yvo Riezebos Design
Photo Credits:	See page C-1

Library of Congress Cataloging-in-Publication Data
Knight, Randall Dewey.
Physics for scientists and engineers : a strategic approach / randall d. knight. -- 3rd ed.
   p. cm.
Includes bibliographical references and index.
ISBN 978-0-321-74090-8
1. Physics--Textbooks. I. Title.
QC23.2.K654 2012
530--dc23
2011033849
ISBN-13: 978-0-321-75317-5 ISBN-10: 0-321-75317-8 (Volume 3)
ISBN-13: 978-0-321-74090-8 ISBN-10: 0-321-74090-4 (Student Edition)
ISBN-13: 978-0-321-76519-2 ISBN-10: 0-321-76519-2 (Instructor's Review Copy)
ISBN-13: 978-0-132-83212-0 ISBN-10: 0-132-83212-7 (NASTA Edition)

1 2 3 4 5 6 7 8 9 10—DOW—15 14 13 12 11

About the Author

Randy Knight has taught introductory physics for over 30 years at Ohio State University and California Polytechnic University, where he is currently Professor of Physics. Professor Knight received a bachelor's degree in physics from Washington University in St. Louis and a Ph.D. in physics from the University of California, Berkeley. He was a post-doctoral fellow at the Harvard-Smithsonian Center for Astrophysics before joining the faculty at Ohio State University. It was at Ohio State that he began to learn about the research in physics education that, many years later, led to this book.

Professor Knight's research interests are in the field of lasers and spectroscopy, and he has published over 25 research papers. He also directs the environmental studies program at Cal Poly, where, in addition to introductory physics, he teaches classes on energy, oceanography, and environmental issues. When he's not in the classroom or in front of a computer, you can find Randy hiking, sea kayaking, playing the piano, or spending time with his wife Sally and their seven cats.

Detailed Contents

Volume 1 contains chapters 1-15; Volume 2 contains chapters 16-19; Volume 3 contains chapters 20-24;
Volume 4 contains chapters 25-36; Volume 5 contains chapters 36-42.

PART

V

Waves and Optics

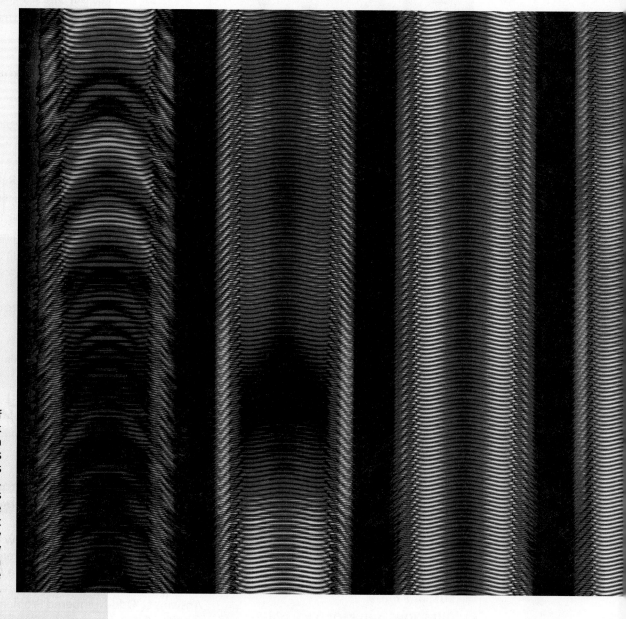

The song of a humpback whale can travel hundreds of kilometers underwater. This graph uses a procedure called wavelet analysis to study the frequency structure of a humpback whale song.

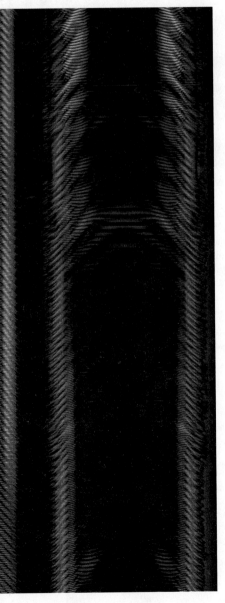

OVERVIEW

The Wave Model

Parts I–IV of this text have been primarily about the physics of particles. You've seen that macroscopic systems ranging from balls and rockets to a gas of molecules can be thought of as particles or as systems of particles. A *particle* is one of the two fundamental models of classical physics. The other, to which we now turn our attention, is a *wave*.

Waves are ubiquitous in nature. Familiar examples of waves include

- Undulating ripples on a pond.
- The swaying ground of an earthquake.
- A vibrating guitar string.
- The sweet sound of a flute.
- The colors of the rainbow.

The physics of waves is the subject of Part V, the next stage of our journey. Despite the great diversity of types and sources of waves, a single, elegant physical theory is capable of describing them all. Our exploration of wave phenomena will call upon sound waves, light waves, and vibrating strings for examples, but our goal is to emphasize the unity and coherence of the ideas that are common to *all* types of waves.

A wave, in contrast with a particle, is diffuse, spread out, not to be found at a single point in space. We will start with waves traveling outward through some medium, like the spreading ripples after a pebble hits a pool of water. These are called *traveling waves.* An investigation of what happens when waves travel through each other will lead us to *standing waves,* which are essential for understanding phenomena ranging from those as common as musical instruments and water sloshing in a tub to as complex as lasers and the electrons in atoms. We'll also study one of the most important defining characteristics of waves—their ability to exhibit *interference.*

Three chapters will be devoted to light and optics, perhaps the most important application of waves. Although light is an electromagnetic wave, your understanding of these chapters will depend on nothing more than the "waviness" of light. You can study these chapters either before or after your study of electricity and magnetism in Part VI. The electromagnetic aspects of light waves will be taken up in Chapter 34.

Our investigation of light will be aided by a second model, the *ray model,* in which light travels in straight lines, reflects from mirrors, and is focused by lenses. Many practical applications of optics, from the camera to the telescope, are best understood with the ray model of light.

In fact, that you're able to read this book at all is due to the first optical instrument you ever used—your eyes. We will investigate the optics of the eye, learn how the cornea and lens form an image on the retina, and see how glasses or contact lenses can be used to correct the image if it is out of focus.

20 Traveling Waves

This surfer is "catching a wave." At the same time, he's seeing light waves and hearing sound waves.

▶ **Looking Ahead** The goal of Chapter 20 is to learn the basic properties of traveling waves.

The Wave Model

A **wave** is a disturbance traveling through a medium. Our goal is to develop a model —the wave model—that describes the basic properties of all waves.

The wave propagates, but the particles of the medium don't. The water molecules simply oscillate up and down as the ripples spread outward.

Two Types of Waves

You'll find that waves come in two basic types:

Transverse waves: The displacement is perpendicular to the direction of travel.

Longitudinal waves: The displacement is parallel to the direction of travel.

Sound and Light

Two types of waves are especially important: sound and light.
- Sound waves are longitudinal waves.
- Light waves are transverse waves.

You'll learn that the colors of visible light correspond to different wavelengths.

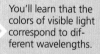

Ultrasound images are made with very-high-frequency sound waves.

Wave Properties

You'll learn that a wave is characterized by three basic properties:
- **Wave speed**: How fast it travels through the medium.
- **Wavelength**: The distance between two neighboring crests.
- **Frequency**: The number of oscillations per second.

You'll also see that wave motion is closely related to simple harmonic motion.

◀ **Looking Back**
Sections 14.1 and 14.2 Properties of simple harmonic motion

Intensity and Loudness

Waves carry energy. The rate at which a wave delivers energy to a surface is the **intensity** of the wave.

Your ears are sensitive to a remarkable range of intensities. You'll learn to use the logarithmic **decibel** scale to characterize the loudness of a sound.

Focusing the sun's light into a smaller area increases its intensity.

The Doppler Effect

The frequency and wavelength of a wave are shifted when there is relative motion between the source and the observer of the waves. This is called the **Doppler effect.**

The pitch of the ambulance siren drops as it races past you. The frequency is shifted up as it approaches, then shifted down as it recedes.

20.1 The Wave Model

Balls, cars, and rockets obviously differ from one another, but the general features of their motions are well described by the *particle model* of Parts I–IV. In Part V we will explore the basic properties of waves with a **wave model,** emphasizing those aspects of wave behavior common to all waves. Although water waves, sound waves, and light waves are clearly different, the wave model will allow us to understand many of the important features they have in common.

The wave model is built around the idea of a **traveling wave,** which is an organized disturbance traveling with a well-defined wave speed. We'll begin our study of traveling waves by looking at two distinct wave motions.

Two types of traveling waves

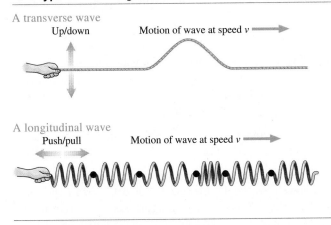

A transverse wave
Up/down
Motion of wave at speed v

A **transverse wave** is a wave in which the displacement is *perpendicular* to the direction in which the wave travels. For example, a wave travels along a string in a horizontal direction while the particles that make up the string oscillate vertically. Electromagnetic waves are also transverse waves because the electromagnetic fields oscillate perpendicular to the direction in which the wave travels.

A longitudinal wave
Push/pull
Motion of wave at speed v

In a **longitudinal wave,** the particles in the medium move *parallel* to the direction in which the wave travels. Here we see a chain of masses connected by springs. If you give the first mass in the chain a sharp push, a disturbance travels down the chain by compressing and expanding the springs. Sound waves in gases and liquids are the most well known examples of longitudinal waves.

We can also classify waves on the basis of what is "waving":

1. **Mechanical waves** travel only within a material *medium*, such as air or water. Two familiar mechanical waves are sound waves and water waves.
2. **Electromagnetic waves,** from radio waves to visible light to x rays, are a self-sustaining oscillation of the *electromagnetic field*. Electromagnetic waves require no material medium and can travel through a vacuum.

The **medium** of a mechanical wave is the substance through or along which the wave moves. For example, the medium of a water wave is the water, the medium of a sound wave is the air, and the medium of a wave on a stretched string is the string. A medium must be *elastic*. That is, a restoring force of some sort brings the medium back to equilibrium after it has been displaced or disturbed. The tension in a stretched string pulls the string back straight after you pluck it. Gravity restores the level surface of a lake after the wave generated by a boat has passed by.

As a wave passes through a medium, the atoms of the medium—we'll simply call them the particles of the medium—are displaced from equilibrium. This is a **disturbance** of the medium. The water ripples of FIGURE 20.1 are a disturbance of the water's surface. A pulse traveling down a string is a disturbance, as are the wake of a boat and the sonic boom created by a jet traveling faster than the speed of sound. **The disturbance of a wave is an *organized* motion of the particles in the medium,** in contrast to the *random* molecular motions of thermal energy.

FIGURE 20.1 Ripples on a pond are a traveling wave.

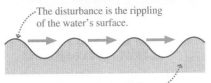

The disturbance is the rippling of the water's surface.

The water is the medium.

Wave Speed

A wave disturbance is created by a *source*. The source of a wave might be a rock thrown into water, your hand plucking a stretched string, or an oscillating loudspeaker cone pushing on the air. Once created, the disturbance travels outward through the medium at the **wave speed** v. This is the speed with which a ripple moves across the water or a pulse travels down a string.

NOTE ▶ The disturbance propagates through the medium, but **the medium as a whole does not move!** The ripples on the pond (the disturbance) move outward from the splash of the rock, but there is no outward flow of water from the splash. Likewise, the particles of a string oscillate up and down but do not move in the direction of a pulse traveling along the string. **A wave transfers energy, but it does not transfer any material or substance outward from the source.** ◀

As an example, we'll prove in Section 20.3 that the wave speed on a string stretched with tension T_s is

$$v_{string} = \sqrt{\frac{T_s}{\mu}} \quad \text{(wave speed on a stretched string)} \quad (20.1)$$

where μ is the string's **linear density,** its mass-to-length ratio:

$$\mu = \frac{m}{L} \quad (20.2)$$

The SI unit of linear density is kg/m. A fat string has a larger value of μ than a skinny string made of the same material. Similarly, a steel wire has a larger value of μ than a plastic string of the same diameter. We'll assume that strings are *uniform,* meaning the linear density is the same everywhere along the length of the string.

NOTE ▶ The subscript s on the symbol T_s for the string's tension distinguishes it from the symbol T for the *period* of oscillation. ◀

Equation 20.1 is the wave *speed,* not the wave velocity, so v_{string} always has a positive value. Every point on a wave travels with this speed. You can increase the wave speed either by *increasing* the string's tension (make it tighter) or by *decreasing* the string's linear density (make it skinnier). We'll examine the implications for stringed musical instruments in Chapter 21.

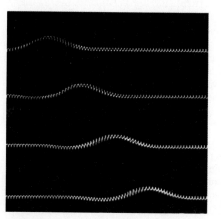

This sequence of photographs shows a wave pulse traveling along a spring.

EXAMPLE 20.1 **Measuring the linear density**

In a laboratory experiment, one end of a metal wire is connected to a motion sensor. The wire is stretched horizontally to a pulley 1.50 m away, then attached to a hanging mass that provides tension. A mechanical pick plucks the horizontal segment of the wire right at the pulley, creating a small wave pulse that travels along the wire. The plucking motion starts a timer that is stopped by the motion sensor when the pulse reaches the end of the wire. Changing the hanging mass changes the time required for the pulse to travel the length of the wire. The data are as follows:

Mass (kg)	Time (ms)
0.50	31
1.00	23
1.50	18
2.00	15
2.50	14

Use the data to determine the wire's linear density.

MODEL The wave pulse is a traveling wave on a stretched string. The hanging mass is in static equilibrium.

VISUALIZE **FIGURE 20.2** is a pictorial representation.

SOLVE The wave speed on the wire is determined by the wire's linear density μ and tension T_s. The hanging mass is in static

FIGURE 20.2 A wave pulse on the wire.

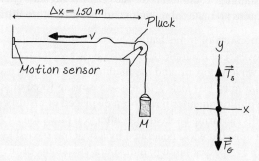

equilibrium, with no net force, so we see from the free-body diagram that the tension in the wire is $T_s = F_G = Mg$. Squaring both sides of Equation 20.1 gives

$$v^2 = \left(\frac{\Delta x}{\Delta t}\right)^2 = \frac{T_s}{\mu} = \frac{Mg}{\mu}$$

Mass M is the independent variable that we've changed, each time measuring the pulse travel time Δt, so we can rearrange the wave-speed equation as

$$(\Delta t)^2 = \frac{\mu (\Delta x)^2}{g} \frac{1}{M}$$

Theory predicts that a graph of the *square* of the travel time versus the *inverse* of the hanging mass should be a straight line passing through the origin with slope $\mu(\Delta x)^2/g$. The graph of FIGURE 20.3, with the times converted from ms to s, is indeed linear with a *y*-intercept of zero. The slope of the best-fit line is seen to be $4.85 \times 10^{-4}\,\mathrm{kg\,s^2}$ (recall that spreadsheets and graphing calculators display this as 4.85E–04), from which we find the wire's linear density:

$$\mu = \frac{g \times \text{slope}}{(\Delta x)^2} = 0.0021\ \mathrm{kg/m} = 2.1\ \mathrm{g/m}$$

ASSESS A meter of thin wire is likely to have a mass of a few grams, so a linear density of a few g/m seems reasonable.

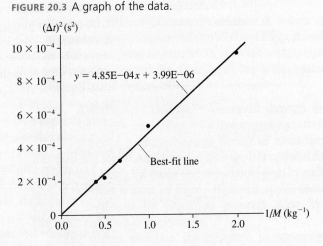

FIGURE 20.3 A graph of the data.

$y = 4.85\mathrm{E}{-}04x + 3.99\mathrm{E}{-}06$

Best-fit line

The wave speed on a string is a property of the string—its tension and linear density. In general, **the wave speed is a property** *of the medium.* The wave speed depends on the restoring forces within the medium but not at all on the shape or size of the pulse, how the pulse was generated, or how far it has traveled.

STOP TO THINK 20.1 Which of the following actions would make a pulse travel faster along a stretched string? More than one answer may be correct. If so, give all that are correct.

a. Move your hand up and down more quickly as you generate the pulse.
b. Move your hand up and down a larger distance as you generate the pulse.
c. Use a heavier string of the same length, under the same tension.
d. Use a lighter string of the same length, under the same tension.
e. Stretch the string tighter to increase the tension.
f. Loosen the string to decrease the tension.
g. Put more force into the wave.

20.2 One-Dimensional Waves

To understand waves we must deal with functions of *two* variables. Until now, we have been concerned with quantities that depend only on time, such as $x(t)$ or $v(t)$. Functions of the one variable *t* are all right for a particle because a particle is only in one place at a time, but a wave is not localized. It is spread out through space at each instant of time. To describe a wave mathematically requires a function that specifies not only an instant of time (when) but also a point in space (where).

Rather than leaping into mathematics, we will start by thinking about waves graphically. Consider the wave pulse shown moving along a stretched string in FIGURE 20.4. (We will consider somewhat artificial triangular and square-shaped pulses in this section to make clear where the edges of the pulse are.) The graph shows the string's displacement Δy at a particular instant of time t_1 as a function of position *x* along the string. This is a "snapshot" of the wave, much like what you might make with a camera whose shutter is opened briefly at t_1. A graph that shows the wave's displacement as a function of position at a single instant of time is called a **snapshot graph.** For a wave on a string, a snapshot graph is literally a picture of the wave at this instant.

FIGURE 20.5 shows a sequence of snapshot graphs as the wave of Figure 20.4 continues to move. These are like successive frames from a movie. Notice that the wave

FIGURE 20.4 A snapshot graph of a wave pulse on a string.

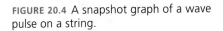

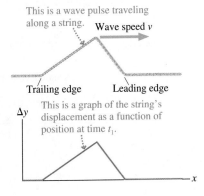

This is a wave pulse traveling along a string. Wave speed *v*

Trailing edge   Leading edge

Δy This is a graph of the string's displacement as a function of position at time t_1.

FIGURE 20.5 A sequence of snapshot graphs shows the wave in motion.

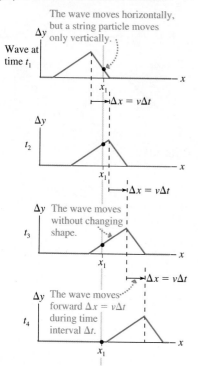

pulse moves forward distance $\Delta x = v\Delta t$ during the time interval Δt. That is, the wave moves with constant speed.

A snapshot graph tells only half the story. It tells us *where* the wave is and how it varies with position, but only at one instant of time. It gives us no information about how the wave *changes* with time. As a different way of portraying the wave, suppose we follow the dot marked on the string in Figure 20.5 and produce a graph showing how the displacement of this dot changes with time. The result, shown in FIGURE 20.6, is a displacement-versus-time graph at a single position in space. A graph that shows the wave's displacement as a function of time at a single position in space is called a **history graph.** It tells the history of that particular point in the medium.

You might think we have made a mistake; the graph of Figure 20.6 is reversed compared to Figure 20.5. It is not a mistake, but it requires careful thought to see why. As the wave moves toward the dot, the steep *leading edge* causes the dot to rise quickly. On the displacement-versus-time graph, *earlier* times (smaller values of t) are to the *left* and later times (larger t) to the right. Thus the leading edge of the wave is on the *left* side of the Figure 20.6 history graph. As you move to the right on Figure 20.6 you see the slowly falling *trailing edge* of the wave as it moves past the dot at later times.

The snapshot graph of Figure 20.4 and the history graph of Figure 20.6 portray complementary information. The snapshot graph tells us how things look throughout all of space, but at only one instant of time. The history graph tells us how things look at all times, but at only one position in space. We need them both to have the full story of the wave. An alternative representation of the wave is the series of graphs in FIGURE 20.7, where we can get a clearer sense of the wave moving forward. But graphs like these are essentially impossible to draw by hand, so it is necessary to move back and forth between snapshot graphs and history graphs.

FIGURE 20.6 A history graph for the dot on the string in Figure 20.5.

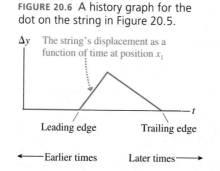

FIGURE 20.7 An alternative look at a traveling wave.

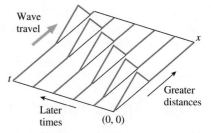

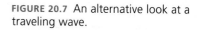

EXAMPLE 20.2 **Finding a history graph from a snapshot graph**

FIGURE 20.8 is a snapshot graph at $t = 0$ s of a wave moving to the right at a speed of 2.0 m/s. Draw a history graph for the position $x = 8.0$ m.

FIGURE 20.8 A snapshot graph at $t = 0$ s.

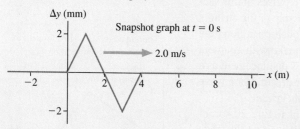

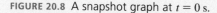

MODEL This is a wave traveling at constant speed. The pulse moves 2.0 m to the right every second.

VISUALIZE The snapshot graph of Figure 20.8 shows the wave at all points on the x-axis at $t = 0$ s. You can see that nothing is happening at $x = 8.0$ m at this instant of time because the wave has not yet reached $x = 8.0$ m. In fact, at $t = 0$ s the leading edge of the wave is still 4.0 m away from $x = 8.0$ m. Because the wave is traveling at 2.0 m/s, it will take 2.0 s for the leading edge to reach $x = 8.0$ m. Thus the history graph for $x = 8.0$ m will be zero until $t = 2.0$ s. The first part of the wave causes a *downward* displacement of the medium, so immediately after $t = 2.0$ s the displacement at $x = 8.0$ m will be negative. The negative portion of the wave pulse is 2.0 m wide and takes 1.0 s to pass $x = 8.0$ m, so the midpoint of the pulse reaches $x = 8.0$ m at $t = 3.0$ s. The positive portion takes another 1.0 s to go past, so the trailing edge of the pulse arrives at $t = 4.0$ s. You could also note that the trailing edge was initially 8.0 m away from $x = 8.0$ m and needed 4.0 s to travel that distance at 2.0 m/s. The displacement at $x = 8.0$ m returns to zero at $t = 4.0$ s and remains zero for all later times. This information is all portrayed on the history graph of FIGURE 20.9.

FIGURE 20.9 The corresponding history graph at $x = 8.0$ m.

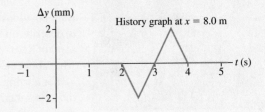

STOP TO THINK 20.2 The graph at the right is the history graph at $x = 4.0$ m of a wave traveling to the right at a speed of 2.0 m/s. Which is the history graph of this wave at $x = 0$ m?

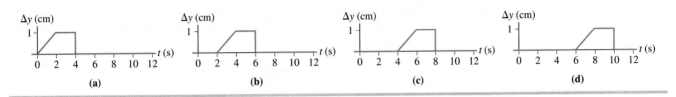

Longitudinal Waves

For a wave on a string, a transverse wave, the snapshot graph is literally a picture of the wave. Not so for a longitudinal wave, where the particles in the medium are displaced parallel to the direction in which the wave is traveling. Thus the displacement is Δx rather than Δy, and a snapshot graph is a graph of Δx versus x.

FIGURE 20.10a is a snapshot graph of a longitudinal wave, such as a sound wave. It's purposefully drawn to have the same shape as the string wave in Example 20.2. Without practice, it's not clear what this graph tells us about the particles in the medium.

FIGURE 20.10 Visualizing a longitudinal wave.

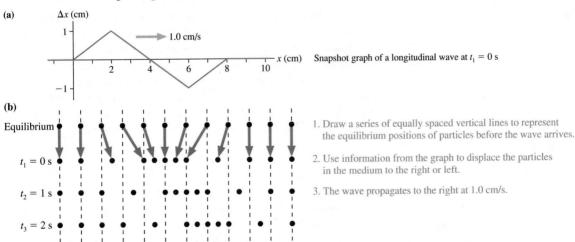

1. Draw a series of equally spaced vertical lines to represent the equilibrium positions of particles before the wave arrives.

2. Use information from the graph to displace the particles in the medium to the right or left.

3. The wave propagates to the right at 1.0 cm/s.

To help you find out, FIGURE 20.10b provides a tool for visualizing longitudinal waves. In the second row, we've used information from the graph to displace the particles in the medium to the right or to the left of their equilibrium positions. For example, the particle at $x = 1.0$ cm has been displaced 0.5 cm to the right because the snapshot graph shows $\Delta x = 0.5$ cm at $x = 1.0$ cm. We now have a picture of the longitudinal wave pulse at $t_1 = 0$ s. You can see that the medium is compressed to higher density at the center of the pulse and, to compensate, expanded to lower density at the leading and trailing edges. Two more lines show the medium at $t_2 = 1$ s and $t_3 = 2$ s so that you can see the wave propagating through the medium at 1.0 cm/s.

You've probably seen or participated in "the wave" at a sporting event. The wave moves around the stadium, but the people (the medium) simply undergo small displacements from their equilibrium positions.

The Displacement

A traveling wave causes the particles of the medium to be displaced from their equilibrium positions. Because one of our goals is to develop a mathematical representation to describe all types of waves, we'll use the generic symbol D to stand for the *displacement* of a wave of any type. But what do we mean by a "particle" in the medium? And what about electromagnetic waves, for which there is no medium?

For a string, where the atoms stay fixed relative to each other, you can think of either the atoms themselves or very small segments of the string as being the particles of the medium. D is then the perpendicular displacement Δy of a point on the string. For a sound wave, D is the longitudinal displacement Δx of a small volume of fluid. For any other mechanical wave, D is the appropriate displacement. Even electromagnetic waves can be described within the same mathematical representation if D is interpreted as a yet-undefined *electromagnetic field strength,* a "displacement" in a more abstract sense as an electromagnetic wave passes through a region of space.

Because the displacement of a particle in the medium depends both on *where* the particle is (position x) and on *when* you observe it (time t), D must be a function of the two variables x and t. That is,

$D(x, t) =$ the displacement at time t of a particle at position x

The values of *both* variables—where and when—must be specified before you can evaluate the displacement D.

20.3 Sinusoidal Waves

A wave source that oscillates with simple harmonic motion (SHM) generates a **sinusoidal wave.** For example, a loudspeaker cone that oscillates in SHM radiates a sinusoidal sound wave. The sinusoidal electromagnetic waves broadcast by television and FM radio stations are generated by electrons oscillating back and forth in the antenna wire with SHM. **The frequency f of the wave is the frequency of the oscillating source.**

FIGURE 20.11 shows a sinusoidal wave moving through a medium. The source of the wave, which is undergoing vertical SHM, is located at $x = 0$. Notice how the wave crests move with steady speed toward larger values of x at later times t.

FIGURE 20.12a is a history graph for a sinusoidal wave, showing the displacement of the medium at one point in space. Each particle in the medium undergoes simple harmonic motion with frequency f, so this graph of SHM is identical to the graphs you learned to work with in Chapter 14. The *period* of the wave, shown on the graph, is the time interval for one cycle of the motion. The period is related to the wave frequency f by

$$T = \frac{1}{f} \tag{20.3}$$

exactly as in simple harmonic motion. The **amplitude** A of the wave is the maximum value of the displacement. The crests of the wave have displacement $D_{\text{crest}} = A$ and the troughs have displacement $D_{\text{trough}} = -A$.

FIGURE 20.11 A sinusoidal wave moving along the x-axis.

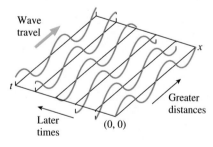

FIGURE 20.12 History and snapshot graphs for a sinusoidal wave.

(a) A history graph at one point in space

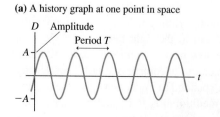

(b) A snapshot graph at one instant of time

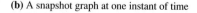
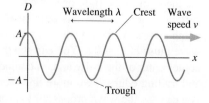

Displacement versus time is only half the story. FIGURE 20.12b shows a snapshot graph for the same wave at one instant in time. Here we see the wave stretched out in space, moving to the right with speed v. An important characteristic of a sinusoidal wave is that it is periodic *in space* as well as in time. As you move from left to right along the "frozen" wave in the snapshot graph, the disturbance repeats itself over and over. The distance spanned by one cycle of the motion is called the **wavelength** of the wave. Wavelength is symbolized by λ (lowercase Greek lambda) and, because it is a length, it is measured in units of meters. The wavelength is shown in Figure 20.12b as the distance between two crests, but it could equally well be the distance between two troughs.

NOTE ▶ Wavelength is the spatial analog of period. The period T is the *time* in which the disturbance at a single point in space repeats itself. The wavelength λ is the *distance* in which the disturbance at one instant of time repeats itself. ◀

The Fundamental Relationship for Sinusoidal Waves

There is an important relationship between the wavelength and the period of a wave. FIGURE 20.13 shows this relationship through five snapshot graphs of a sinusoidal wave at time increments of one-quarter of the period T. One full period has elapsed between the first graph and the last, which you can see by observing the motion at a fixed point on the x-axis. Each point in the medium has undergone exactly one complete oscillation.

The critical observation is that the wave crest marked by an arrow has moved one full wavelength between the first graph and the last. That is, **during a time interval of exactly one period T, each crest of a sinusoidal wave travels forward a distance of exactly one wavelength λ.** Because speed is distance divided by time, the wave speed must be

$$v = \frac{\text{distance}}{\text{time}} = \frac{\lambda}{T} \tag{20.4}$$

Because $f = 1/T$, it is customary to write Equation 20.4 in the form

$$v = \lambda f \tag{20.5}$$

Although Equation 20.5 has no special name, it is *the* fundamental relationship for periodic waves. When using it, keep in mind the *physical* meaning that **a wave moves forward a distance of one wavelength during a time interval of one period.**

NOTE ▶ Wavelength and period are defined only for *periodic* waves, so Equations 20.4 and 20.5 apply only to periodic waves. A wave pulse has a wave speed, but it doesn't have a wavelength or a period. Hence Equations 20.4 and 20.5 cannot be applied to wave pulses. ◀

Because the wave speed is a property of the medium while the wave frequency is a property of the source, it is often useful to write Equation 20.5 as

$$\lambda = \frac{v}{f} = \frac{\text{property of the medium}}{\text{property of the source}} \tag{20.6}$$

The wavelength is a *consequence* of a wave of frequency f traveling through a medium in which the wave speed is v.

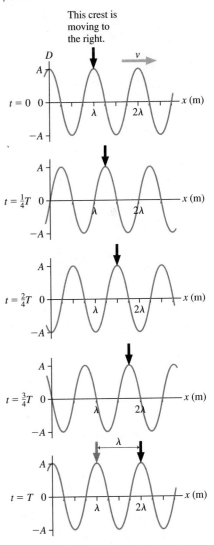

FIGURE 20.13 A series of snapshot graphs at time increments of one-quarter of the period T.

During a time interval of exactly one period, the crest has moved forward exactly one wavelength.

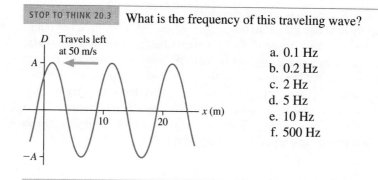

STOP TO THINK 20.3 What is the frequency of this traveling wave?

a. 0.1 Hz
b. 0.2 Hz
c. 2 Hz
d. 5 Hz
e. 10 Hz
f. 500 Hz

The Mathematics of Sinusoidal Waves

FIGURE 20.14 A sinusoidal wave is "frozen" at $t = 0$.

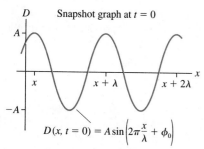

FIGURE 20.14 shows a snapshot graph at $t = 0$ of a sinusoidal wave. The sinusoidal function that describes the displacement of this wave is

$$D(x, t = 0) = A \sin\left(2\pi \frac{x}{\lambda} + \phi_0\right) \qquad (20.7)$$

where the notation $D(x, t = 0)$ means that we've frozen the time at $t = 0$ to make the displacement a function of only x. The term ϕ_0 is a *phase constant* that characterizes the initial conditions. (We'll return to the phase constant momentarily.)

The function of Equation 20.7 is periodic with period λ. We can see this by writing

$$D(x + \lambda) = A \sin\left(2\pi \frac{(x + \lambda)}{\lambda} + \phi_0\right) = A \sin\left(2\pi \frac{x}{\lambda} + \phi_0 + 2\pi \text{ rad}\right)$$

$$= A \sin\left(2\pi \frac{x}{\lambda} + \phi_0\right) = D(x)$$

where we used the fact that $\sin(a + 2\pi \text{ rad}) = \sin a$. In other words, the disturbance created by the wave at $x + \lambda$ is exactly the same as the disturbance at x.

The next step—and it's an important step to graph—is to set the wave in motion. We can do this by replacing x in Equation 20.7 with $x - vt$. To see why this works, recall that the wave moves distance vt during time t. In other words, whatever displacement the wave has at position x at time t, the wave must have had that same displacement at position $x - vt$ at the earlier time $t = 0$. Mathematically, this idea can be captured by writing

$$D(x, t) = D(x - vt, t = 0) \qquad (20.8)$$

Make sure you understand how this statement describes a wave moving in the positive x-direction at speed v.

This is what we were looking for. $D(x, t)$ is the general function describing the traveling wave. It's found by taking the function that describes the wave at $t = 0$—the function of Equation 20.7—and replacing x with $x - vt$. Thus the displacement equation of a sinusoidal wave traveling in the positive x-direction at speed v is

$$D(x, t) = A \sin\left(2\pi \frac{x - vt}{\lambda} + \phi_0\right) = A \sin\left(2\pi \left(\frac{x}{\lambda} - \frac{t}{T}\right) + \phi_0\right) \qquad (20.9)$$

In the last step we used $v = \lambda f = \lambda/T$ to write $v/\lambda = 1/T$. The function of Equation 20.9 is not only periodic in space with period λ, it is also periodic in time with period T. That is, $D(x, t + T) = D(x, t)$.

It will be useful to introduce two new quantities. First, recall from simple harmonic motion the *angular frequency*

$$\omega = 2\pi f = \frac{2\pi}{T} \tag{20.10}$$

The units of ω are rad/s, although many textbooks use simply s^{-1}.

You can see that ω is 2π times the reciprocal of the period in time. This suggests that we define an analogous quantity, called the **wave number** k, that is 2π times the reciprocal of the period in space:

$$k = \frac{2\pi}{\lambda} \tag{20.11}$$

The units of k are rad/m, although many textbooks use simply m^{-1}.

NOTE ▶ The wave number k is *not* a spring constant, even though it uses the same symbol. This is a most unfortunate use of symbols, but every major textbook and professional tradition uses the same symbol k for these two very different meanings, so we have little choice but to follow along. ◀

We can use the fundamental relationship $v = \lambda f$ to find an analogous relationship between ω and k:

$$v = \lambda f = \frac{2\pi}{k}\frac{\omega}{2\pi} = \frac{\omega}{k} \tag{20.12}$$

which is usually written

$$\omega = vk \tag{20.13}$$

Equation 20.13 contains no new information. It is a variation of Equation 20.5, but one that is convenient when working with k and ω.

If we use the definitions of Equations 20.10 and 20.11, Equation 20.9 for the displacement can be written

$$D(x, t) = A\sin(kx - \omega t + \phi_0) \tag{20.14}$$
(sinusoidal wave traveling in the positive *x*-direction)

A sinusoidal wave traveling in the negative *x*-direction is $A\sin(kx + \omega t + \phi_0)$. Equation 20.14 is graphed versus x and t in FIGURE 20.15.

Just as it did for simple harmonic motion, the phase constant ϕ_0 characterizes the initial conditions. At $(x, t) = (0 \text{ m}, 0 \text{ s})$ Equation 20.14 becomes

$$D(0 \text{ m}, 0 \text{ s}) = A\sin\phi \tag{20.15}$$

Different values of ϕ_0 describe different initial conditions for the wave.

FIGURE 20.15 Interpreting the equation of a sinusoidal traveling wave.

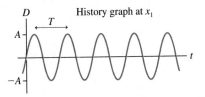

If x is fixed, $D(x_1, t) = A\sin(kx_1 - \omega t + \phi_0)$ gives a sinusoidal history graph at one point in space, x_1. It repeats every T s.

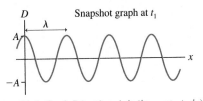

If t is fixed, $D(x, t_1) = A\sin(kx - \omega t_1 + \phi_0)$ gives a sinusoidal snapshot graph at one instant of time, t_1. It repeats every λ m.

EXAMPLE 20.3 **Analyzing a sinusoidal wave**

A sinusoidal wave with an amplitude of 1.00 cm and a frequency of 100 Hz travels at 200 m/s in the positive *x*-direction. At $t = 0$ s, the point $x = 1.00$ m is on a crest of the wave.

a. Determine the values of A, v, λ, k, f, ω, T, and ϕ_0 for this wave.
b. Write the equation for the wave's displacement as it travels.
c. Draw a snapshot graph of the wave at $t = 0$ s.

VISUALIZE The snapshot graph will be sinusoidal, but we must do some numerical analysis before we know how to draw it.

SOLVE a. There are several numerical values associated with a sinusoidal traveling wave, but they are not all independent. From the problem statement itself we learn that

$$A = 1.00 \text{ cm} \quad v = 200 \text{ m/s} \quad f = 100 \text{ Hz}$$

We can then find:

$$\lambda = v/f = 2.00 \text{ m}$$

$$k = 2\pi/\lambda = \pi \text{ rad/m or } 3.14 \text{ rad/m}$$

Continued

$$\omega = 2\pi f = 628 \text{ rad/s}$$

$$T = 1/f = 0.0100 \text{ s} = 10.0 \text{ ms}$$

The phase constant ϕ_0 is determined by the initial conditions. We know that a wave crest, with displacement $D = A$, is passing $x_0 = 1.00$ m at $t_0 = 0$ s. Equation 20.14 at x_0 and t_0 is

$$D(x_0, t_0) = A = A \sin\left(k(1.00 \text{ m}) + \phi_0\right)$$

This equation is true only if $\sin\left(k(1.00 \text{ m}) + \phi_0\right) = 1$, which requires

$$k(1.00 \text{ m}) + \phi_0 = \frac{\pi}{2} \text{ rad}$$

Solving for the phase constant gives

$$\phi_0 = \frac{\pi}{2} \text{ rad} - (\pi \text{ rad/m})(1.00 \text{ m}) = -\frac{\pi}{2} \text{ rad}$$

b. With the information gleaned from part a, the wave's displacement is

$$D(x, t) = 1.00 \text{ cm} \times$$

$$\sin\left[(3.14 \text{ rad/m})x - (628 \text{ rad/s})t - \pi/2 \text{ rad}\right]$$

Notice that we included units with A, k, ω, and ϕ_0.

c. We know that $x = 1.00$ m is a wave crest at $t = 0$ s and that the wavelength is $\lambda = 2.00$ m. Because the origin is $\lambda/2$ away from the crest at $x = 1.00$ m, we expect to find a wave trough at $x = 0$. This is confirmed by calculating $D(0 \text{ m}, 0 \text{ s}) = (1.00 \text{ cm}) \sin(-\pi/2 \text{ rad}) = -1.00$ cm. FIGURE 20.16 is a snapshot graph that portrays this information.

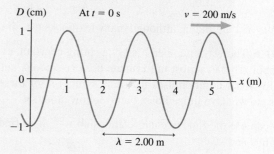

FIGURE 20.16 A snapshot graph at $t = 0$ s of the sinusoidal wave of Example 20.3.

Wave Motion on a String

The displacement equation, Equation 20.14, allows us to learn more about wave motion on a string. As a wave travels along the x-axis, the points on the string oscillate back and forth in the y-direction. The displacement D of a point on the string is simply that point's y-coordinate, so Equation 20.14 for a string wave is

$$y(x, t) = A \sin(kx - \omega t + \phi_0) \tag{20.16}$$

The velocity of a particle on the string—**which is not the same as the velocity of the wave along the string**—is the time derivative of Equation 20.16:

$$v_y = \frac{dy}{dt} = -\omega A \cos(kx - \omega t + \phi_0) \tag{20.17}$$

The maximum velocity of a small segment of the string is $v_{max} = \omega A$. This is the same result we found for simple harmonic motion because the motion of the string particles is simple harmonic motion. FIGURE 20.17 shows velocity vectors *of the particles* at different points on a sinusoidal wave.

NOTE ▶ Creating a wave of larger amplitude increases the speed of particles in the medium, but it does *not* change the speed of the wave *through* the medium. ◀

Pursuing this line of thought, we can derive an expression for the wave speed along the string. FIGURE 20.18 shows a small segment of the string with length $\Delta x \ll \lambda$ right at a crest of the wave. You can see that the string's tension exerts a downward force on this piece of the string, pulling it back to equilibrium. Newton's second law for this small segment of string is

$$(F_{net})_y = ma_y = (\mu \Delta x) a_y \tag{20.18}$$

where we used the string's linear density μ to write the mass as $m = \mu \Delta x$.

FIGURE 20.17 A snapshot graph of a wave on a string with vectors showing the velocity *of the string* at various points.

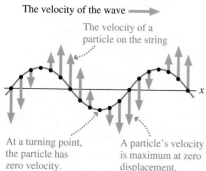

The velocity of the wave ⟶

The velocity of a particle on the string

At a turning point, the particle has zero velocity.

A particle's velocity is maximum at zero displacement.

From simple harmonic motion, we know that this point of maximum displacement is also the point of maximum acceleration. The acceleration of a point on the string is the time derivative of Equation 20.17:

$$a_y = \frac{dv_y}{dt} = -\omega^2 A \sin(kx - \omega t + \phi_0) \qquad (20.19)$$

FIGURE 20.18 A small segment of string at the crest of a wave.

A small segment of the string at the crest of the wave. Because of the curvature of the string, the tension forces exert a net downward force on this segment.

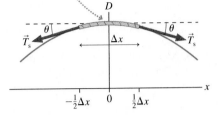

Thus the acceleration at the crest of the wave is $a_y = -\omega^2 A$. But the angular frequency ω with which the particles of the string oscillate is related to the wave's speed v along the string by Equation 20.13, $\omega = vk$. Thus

$$a_y = -\omega^2 A = -v^2 k^2 A \qquad (20.20)$$

A large wave speed causes the particles of the string to oscillate more quickly and thus to have a larger acceleration.

You can see from Figure 20.18 that the y-component of the tension is $T_s \sin\theta$, where θ is the angle of the string at $x = \frac{1}{2}\Delta x$. θ is a *negative* angle because it is below the x-axis. This segment of string is pulled from both ends, so

$$(F_{net})_y = 2T_s \sin\theta \qquad (20.21)$$

The angle θ is very small because $\Delta x \ll \lambda$, so we can use the small-angle approximation ($\sin u \approx \tan u$ if $u \ll 1$) to write

$$(F_{net})_y \approx 2T_s \tan\theta \qquad (20.22)$$

where $\tan\theta$ is the slope of the string at $x = \frac{1}{2}\Delta x$.

At this specific instant, with the crest of the wave at $x = 0$, the equation of the string is

$$y = A\cos(kx)$$

The slope of the string at $x = \frac{1}{2}\Delta x$ is the derivative evaluated at that point:

$$\tan\theta = \frac{dy}{dx}\Big|_{\text{at } \Delta x/2} = -kA\sin(kx)\big|_{\text{at } \Delta x/2} = -kA\sin\left(\frac{k\Delta x}{2}\right)$$

Now $\Delta x \ll \lambda$, so $k\Delta x/2 = \pi\Delta x/\lambda \ll 1$. Thus the small-angle approximation ($\sin u \approx u$ if $u \ll 1$) of the slope is

$$\tan\theta \approx -kA\left(\frac{k\Delta x}{2}\right) = -\frac{k^2 A\Delta x}{2} \qquad (20.23)$$

If we substitute this expression for $\tan\theta$ into Equation 20.22, we find that the net force on this little piece of string is

$$(F_{net})_y = -k^2 A T_s \Delta x \qquad (20.24)$$

Now we can use Equation 20.20 for a_y and Equation 20.24 for $(F_{net})_y$ in Newton's second law. With these substitutions, Equation 20.18 becomes

$$(F_{net})_y = -k^2 A T_s \Delta x = (\mu\Delta x)a_y = -v^2 k^2 A\mu\Delta x \qquad (20.25)$$

The term $-k^2 A\Delta x$ cancels, and we're left with

$$v = \sqrt{\frac{T_s}{\mu}} \qquad (20.26)$$

This was the result that we stated, without proof, in Equation 20.1. Although we've derived Equation 20.26 with the assumption of a sinusoidal wave, the wave speed does not depend on the shape of the wave. Thus any wave on a stretched string will have this wave speed.

EXAMPLE 20.4 **Generating a sinusoidal wave**

A very long string with $\mu = 2.0$ g/m is stretched along the x-axis with a tension of 5.0 N. At $x = 0$ m it is tied to a 100 Hz simple harmonic oscillator that vibrates perpendicular to the string with an amplitude of 2.0 mm. The oscillator is at its maximum positive displacement at $t = 0$ s.

a. Write the displacement equation for the traveling wave on the string.
b. At $t = 5.0$ ms, what is the string's displacement at a point 2.7 m from the oscillator?

MODEL The oscillator generates a sinusoidal traveling wave on a string. The displacement of the wave has to match the displacement of the oscillator at $x = 0$ m.

SOLVE a. The equation for the displacement is

$$D(x, t) = A \sin(kx - \omega t + \phi_0)$$

with A, k, ω, and ϕ_0 to be determined. The wave amplitude is the same as the amplitude of the oscillator that generates the wave, so $A = 2.0$ mm. The oscillator has its maximum displacement $y_{osc} = A = 2.0$ mm at $t = 0$ s, thus

$$D(0 \text{ m}, 0 \text{ s}) = A \sin(\phi_0) = A$$

This requires the phase constant to be $\phi_0 = \pi/2$ rad. The wave's frequency is $f = 100$ Hz, the frequency of the source;

therefore the angular frequency is $\omega = 2\pi f = 200\pi$ rad/s. We still need $k = 2\pi/\lambda$, but we do not know the wavelength. However, we have enough information to determine the wave speed, and we can then use either $\lambda = v/f$ or $k = \omega/v$. The speed is

$$v = \sqrt{\frac{T_s}{\mu}} = \sqrt{\frac{5.0 \text{ N}}{0.0020 \text{ kg/m}}} = 50 \text{ m/s}$$

Using v, we find $\lambda = 0.50$ m and $k = 2\pi/\lambda = 4\pi$ rad/m. Thus the wave's displacement equation is

$$D(x, t) = (2.0 \text{ mm}) \times$$
$$\sin\left[2\pi\big((2.0 \text{ m}^{-1})x - (100 \text{ s}^{-1})t\big) + \pi/2 \text{ rad} \right]$$

Notice that we have separated out the 2π. This step is not essential, but for some problems it makes subsequent steps easier.

b. The wave's displacement at $t = 5.0$ ms $= 0.0050$ s is

$$D(x, t = 5.0 \text{ ms}) = (2.0 \text{ mm}) \sin(4\pi x - \pi \text{ rad} + \pi/2 \text{ rad})$$
$$= (2.0 \text{ mm}) \sin(4\pi x - \pi/2 \text{ rad})$$

At $x = 2.7$ m (calculator set to radians!), the displacement is

$$D(2.7 \text{ m}, 5.0 \text{ ms}) = 1.6 \text{ mm}$$

20.4 Waves in Two and Three Dimensions

Suppose you were to take a photograph of ripples spreading on a pond. If you mark the location of the *crests* on the photo, your picture would look like FIGURE 20.19a. The lines that locate the crests are called **wave fronts,** and they are spaced precisely one wavelength apart. The diagram shows only a single instant of time, but you can imagine a movie in which you would see the wave fronts moving outward from the source at speed v. A wave like this is called a **circular wave.** It is a two-dimensional wave that spreads across a surface.

Although the wave fronts are circles, you would hardly notice the curvature if you observed a small section of the wave front very, very far away from the source. The wave fronts would appear to be parallel lines, still spaced one wavelength apart and traveling at speed v. A good example is an ocean wave reaching a beach. Ocean waves are generated by storms and wind far out at sea, hundreds or thousands of miles away. By the time they reach the beach where you are working on your tan, the crests appear to be straight lines. An aerial view of the ocean would show a wave diagram like FIGURE 20.19b.

Many waves of interest, such as sound waves or light waves, move in three dimensions. For example, loudspeakers and lightbulbs emit **spherical waves.** That is, the crests of the wave form a series of concentric spherical shells separated by the wavelength λ. In essence, the waves are three-dimensional ripples. It will still be useful to draw wave-front diagrams such as Figure 20.19, but now the circles are slices through the spherical shells locating the wave crests.

If you observe a spherical wave very, very far from its source, the small piece of the wave front that you can see is a little patch on the surface of a very large sphere. If the radius of the sphere is sufficiently large, you will not notice the curvature and this little patch of the wave front appears to be a plane. FIGURE 20.20 illustrates the idea of a **plane wave.**

FIGURE 20.19 The wave fronts of a circular or spherical wave.

(a)

Wave fronts are the crests of the wave. They are spaced one wavelength apart.

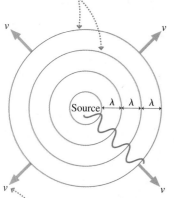

The circular wave fronts move outward from the source at speed v.

(b)

Very far away from the source, small sections of the wave fronts appear to be straight lines.

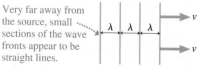

To visualize a plane wave, imagine standing on the x-axis facing a sound wave as it comes toward you from a very distant loudspeaker. Sound is a longitudinal wave, so the particles of medium oscillate toward you and away from you. If you were to locate all of the particles that, at one instant of time, were at their maximum displacement toward you, they would all be located in a plane perpendicular to the travel direction. This is one of the wave fronts in Figure 20.20, and all the particles in this plane are doing exactly the same thing at that instant of time. This plane is moving toward you at speed v. There is another plane one wavelength behind it where the molecules are also at maximum displacement, yet another two wavelengths behind the first, and so on.

Because a plane wave's displacement depends on x but not on y or z, the displacement function $D(x, t)$ describes a plane wave just as readily as it does a one-dimensional wave. Once you specify a value for x, the displacement is the same at every point in the yz-plane that slices the x-axis at that value (i.e., one of the planes shown in Figure 20.20).

NOTE ▶ There are no perfect plane waves in nature, but many waves of practical interest can be modeled as plane waves. ◀

We can describe a circular wave or a spherical wave by changing the mathematical description from $D(x, t)$ to $D(r, t)$, where r is the radial distance measured outward from the source. Then the displacement of the medium will be the same at every point on a spherical surface. In particular, a sinusoidal spherical wave with wave number k and angular frequency ω is written

$$D(r, t) = A(r)\sin(kr - \omega t + \phi_0) \qquad (20.27)$$

Other than the change of x to r, the only difference is that the amplitude is now a function of r. A one-dimensional wave propagates with no change in the wave amplitude. But circular and spherical waves spread out to fill larger and larger volumes of space. To conserve energy, an issue we'll look at later in the chapter, the wave's amplitude has to decrease with increasing distance r. This is why sound and light decrease in intensity as you get farther from the source. We don't need to specify exactly how the amplitude decreases with distance, but you should be aware that it does.

Phase and Phase Difference

The quantity $(kx - \omega t + \phi_0)$ is called the **phase** of the wave, denoted ϕ. The phase of a wave will be an important concept in Chapters 21 and 22, where we will explore the consequences of adding various waves together. For now, we can note that the wave fronts seen in Figures 20.19 and 20.20 are "surfaces of constant phase." To see this, use the phase to write the displacement as simply $D(x, t) = A\sin\phi$. Because each point on a wave front has the same displacement, the phase must be the same at every point.

It will be useful to know the *phase difference* $\Delta\phi$ between two different points on a sinusoidal wave. FIGURE 20.21 shows two points on a sinusoidal wave at time t. The phase difference between these points is

$$\Delta\phi = \phi_2 - \phi_1 = (kx_2 - \omega t + \phi_0) - (kx_1 - \omega t + \phi_0)$$
$$= k(x_2 - x_1) = k\Delta x = 2\pi\frac{\Delta x}{\lambda} \qquad (20.28)$$

That is, **the phase difference between two points on a wave depends on only the ratio of their separation Δx to the wavelength λ.** For example, two points on a wave separated by $\Delta x = \frac{1}{2}\lambda$ have a phase difference $\Delta\phi = \pi$ rad.

An important consequence of Equation 20.28 is that **the phase difference between two adjacent wave fronts is $\Delta\phi = 2\pi$ rad.** This follows from the fact that two adjacent wave fronts are separated by $\Delta x = \lambda$. This is an important idea. Moving from one crest of the wave to the next corresponds to changing the *distance* by λ and changing the *phase* by 2π rad.

FIGURE 20.20 A plane wave.

Very far from the source, small segments of spherical wave fronts appear to be planes. The wave is cresting at every point in these planes.

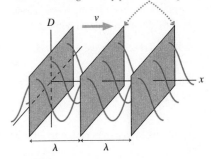

FIGURE 20.21 The phase difference between two points on a wave.

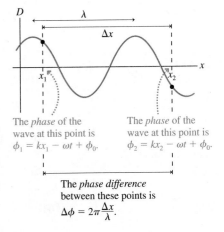

The *phase* of the wave at this point is $\phi_1 = kx_1 - \omega t + \phi_0$.

The *phase* of the wave at this point is $\phi_2 = kx_2 - \omega t + \phi_0$.

The *phase difference* between these points is

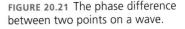

$$\Delta\phi = 2\pi\frac{\Delta x}{\lambda}.$$

EXAMPLE 20.5 **The phase difference between two points on a sound wave**

A 100 Hz sound wave travels with a wave speed of 343 m/s.

a. What is the phase difference between two points 60.0 cm apart along the direction the wave is traveling?

b. How far apart are two points whose phase differs by 90°?

MODEL   Treat the wave as a plane wave traveling in the positive x-direction.

SOLVE   a. The phase difference between two points is

$$\Delta\phi = 2\pi\frac{\Delta x}{\lambda}$$

In this case, $\Delta x = 60.0$ cm $= 0.600$ m. The wavelength is

$$\lambda = \frac{v}{f} = \frac{343 \text{ m/s}}{100 \text{ Hz}} = 3.43 \text{ m}$$

and thus

$$\Delta\phi = 2\pi\frac{0.600 \text{ m}}{3.43 \text{ m}} = 0.350\pi \text{ rad} = 63.0°$$

b. A phase difference $\Delta\phi = 90°$ is $\pi/2$ rad. This will be the phase difference between two points when $\Delta x/\lambda = \frac{1}{4}$, or when $\Delta x = \lambda/4$. Here, with $\lambda = 3.43$ m, $\Delta x = 85.8$ cm.

ASSESS   The phase difference increases as Δx increases, so we expect the answer to part b to be larger than 60 cm.

STOP TO THINK 20.4   What is the phase difference between the crest of a wave and the adjacent trough?

a. -2π rad
b. 0 rad
c. $\pi/4$ rad
d. $\pi/2$ rad
e. π rad
f. 3π rad

FIGURE 20.22 A sound wave in a fluid is a sequence of compressions and rarefactions. The variation in density and the amount of motion have been greatly exaggerated.

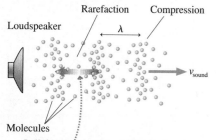

Individual molecules oscillate back and forth with displacement D. As they do so, the compressions propagate forward at speed v_{sound}. Because compressions are regions of higher pressure, a sound wave can be thought of as a pressure wave.

TABLE 20.1 The speed of sound

Medium	Speed (m/s)
Air (0°C)	331
Air (20°C)	343
Helium (0°C)	970
Ethyl alcohol	1170
Water	1480
Granite	6000
Aluminum	6420

20.5 Sound and Light

Although there are many kinds of waves in nature, two are especially significant for us as humans. These are sound waves and light waves, the basis of hearing and seeing.

Sound Waves

We usually think of sound waves traveling in air, but sound can travel through any gas, through liquids, and even through solids. FIGURE 20.22 shows a loudspeaker cone vibrating back and forth in a fluid such as air or water. Each time the cone moves forward, it collides with the molecules and pushes them closer together. A half cycle later, as the cone moves backward, the fluid has room to expand and the density decreases a little. These regions of higher and lower density (and thus higher and lower pressure) are called **compressions** and **rarefactions.**

This periodic sequence of compressions and rarefactions travels outward from the loudspeaker as a longitudinal sound wave. When the wave reaches your ear, the oscillating pressure causes your eardrum to vibrate. These vibrations are transferred into your inner ear and perceived as sound.

Your ears are able to detect sinusoidal sound waves with frequencies between about 20 Hz and about 20,000 Hz, or 20 kHz. Low frequencies are perceived as "low pitch" bass notes, while high frequencies are heard as "high pitch" treble notes. Your high-frequency range of hearing can deteriorate either with age or as a result of exposure to loud sounds that damage the ear.

The speed of sound waves depends on the properties of the medium. A thermodynamic analysis of the compressions and expansions shows that the wave speed in a gas depends on the temperature and on the molecular mass of the gas. For air at room temperature (20°C),

$$v_{sound} = 343 \text{ m/s} \qquad \text{(sound speed in air at 20°C)}$$

The speed of sound is a little lower at lower temperatures and a little higher at higher temperatures. Liquids and solids are less compressible than air, and that makes the speed of sound in those media higher than in air. Table 20.1 gives the speed of sound in several substances.

A speed of 343 m/s is high, but not extraordinarily so. A distance as small as 100 m is enough to notice a slight delay between when you see something, such as a person hammering a nail, and when you hear it. The time required for sound to travel 1 km is $t = (1000 \text{ m})/(343 \text{ m/s}) \approx 3$ s. You may have learned to estimate the distance to a bolt of lightning by timing the number of seconds between when you see the flash and when you hear the thunder. Because sound takes 3 s to travel 1 km, the time divided by 3 gives the distance in kilometers. Or, in English units, the time divided by 5 gives the distance in miles.

Sound waves exist at frequencies well above 20 kHz, even though humans can't hear them. These are called *ultrasonic* frequencies. Oscillators vibrating at frequencies of many MHz generate the ultrasonic waves used in ultrasound medical imaging. A 3 MHz wave traveling through water (which is basically what your body is) at a sound speed of 1480 m/s has a wavelength of about 0.5 mm. It is this very small wavelength that allows ultrasound to image very small objects. We'll see why when we study *diffraction* in Chapter 22.

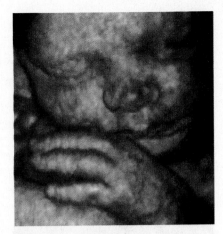

This ultrasound image is an example of using high-frequency sound waves to "see" within the human body.

EXAMPLE 20.6 **Sound wavelengths**

What are the wavelengths of sound waves at the limits of human hearing and at the midrange frequency of 500 Hz? Notes sung by human voices are near 500 Hz, as are notes played by striking keys near the center of a piano keyboard.

MODEL Assume a room temperature of 20°C.

SOLVE We can use the fundamental relationship $\lambda = v/f$ to find the wavelengths for sounds of various frequencies:

$$f = 20 \text{ Hz} \qquad \lambda = \frac{343 \text{ m/s}}{20 \text{ Hz}} = 17 \text{ m}$$

$$f = 500 \text{ Hz} \qquad \lambda = \frac{343 \text{ m/s}}{500 \text{ Hz}} = 0.69 \text{ m}$$

$$f = 20,000 \text{ Hz} \qquad \lambda = \frac{343 \text{ m/s}}{20,000 \text{ Hz}} = 0.017 \text{ m} = 1.7 \text{ cm}$$

ASSESS The wavelength of a 20 kHz note is a small 1.7 cm while, at the other extreme, a 20 Hz note has a huge wavelength of 17 m! This is because a wave moves forward one wavelength during a time interval of one period, and a wave traveling at 343 m/s can move 17 m during the $\frac{1}{20}$ s period of a 20 Hz note. The 69 cm wavelength of a 500 Hz note is more of a "human scale." You might note that most musical instruments are a meter or a little less in size. This is not a coincidence. You will see in the next chapter how the wavelength produced by a musical instrument is related to its size.

Electromagnetic Waves

A light wave is an *electromagnetic wave,* an oscillation of the electromagnetic field. Other electromagnetic waves, such as radio waves, microwaves, and ultraviolet light, have the same physical characteristics as light waves even though we cannot sense them with our eyes. It is easy to demonstrate that light will pass unaffected through a container from which all the air has been removed, and light reaches us from distant stars through the vacuum of interstellar space. Such observations raise interesting but difficult questions. If light can travel through a region in which there is no matter, then what is the *medium* of a light wave? What is it that is waving?

It took scientists over 50 years, most of the 19th century, to answer this question. We will examine the answers in more detail in Part IV after we introduce the ideas of electric and magnetic fields. For now we can say that light waves are a "self-sustaining oscillation of the electromagnetic field." That is, the displacement D is an electric or magnetic field. Being self-sustaining means that electromagnetic waves require *no material medium* in order to travel; hence electromagnetic waves are not mechanical waves. Fortunately, we can learn about the wave properties of light without having to understand electromagnetic fields.

It was predicted theoretically in the late 19th century, and has been subsequently confirmed, that all electromagnetic waves travel through vacuum with the same speed, called the *speed of light.* The value of the speed of light is

$$v_{\text{light}} = c = 299,792,458 \text{ m/s} \qquad \text{(electromagnetic wave speed in vacuum)}$$

where the special symbol c is used to designate the speed of light. (This is the c in Einstein's famous formula $E = mc^2$.) Now *this* is really moving—about one million times faster than the speed of sound in air!

NOTE ▶ $c = 3.00 \times 10^8$ m/s is the appropriate value to use in calculations. ◀

The wavelengths of light are extremely small. You will learn in Chapter 22 how these wavelengths are determined, but for now we will note that visible light is an electromagnetic wave with a wavelength (in air) in the range of roughly 400 nm (400×10^{-9} m) to 700 nm (700×10^{-9} m). Each wavelength is perceived as a different color, with the longer wavelengths seen as orange or red light and the shorter wavelengths seen as blue or violet light. A prism is able to spread the different wavelengths apart, from which we learn that "white light" is all the colors, or wavelengths, combined. The spread of colors seen with a prism, or seen in a rainbow, is called the *visible spectrum.*

If the wavelengths of light are unbelievably small, the oscillation frequencies are unbelievably large. The frequency for a 600 nm wavelength of light (orange) is

$$f = \frac{v}{\lambda} = \frac{3.00 \times 10^8 \text{ m/s}}{600 \times 10^{-9} \text{ m}} = 5.00 \times 10^{14} \text{ Hz}$$

The frequencies of light waves are roughly a factor of a trillion (10^{12}) higher than sound frequencies.

Electromagnetic waves exist at many frequencies other than the rather limited range that our eyes detect. One of the major technological advances of the 20th century was learning to generate and detect electromagnetic waves at many frequencies, ranging from low-frequency radio waves to the extraordinarily high frequencies of x rays. **FIGURE 20.23** shows that the visible spectrum is a small slice of the much broader **electromagnetic spectrum.**

White light passing through a prism is spread out into a band of colors called the *visible spectrum.*

FIGURE 20.23 The electromagnetic spectrum from 10^6 Hz to 10^{18} Hz.

Increasing frequency (Hz) ⟶

10^6	10^8	10^{10}	10^{12}	10^{14}	10^{16}	10^{18}
AM radio	FM radio/TV	Microwaves		Infrared	Ultraviolet	X rays
300	3	0.03	3×10^{-4}	3×10^{-6}	3×10^{-8}	3×10^{-10}

⟵ Increasing wavelength (m)

Visible light

700 nm   600 nm   500 nm   400 nm

EXAMPLE 20.7 **Traveling at the speed of light**

A satellite exploring Jupiter transmits data to the earth as a radio wave with a frequency of 200 MHz. What is the wavelength of the electromagnetic wave, and how long does it take the signal to travel 800 million kilometers from Jupiter to the earth?

SOLVE Radio waves are sinusoidal electromagnetic waves traveling with speed c. Thus

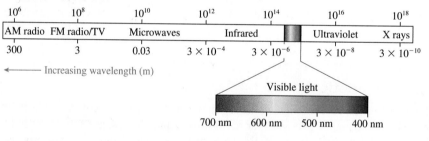

$$\lambda = \frac{c}{f} = \frac{3.00 \times 10^8 \text{ m/s}}{2.00 \times 10^8 \text{ Hz}} = 1.5 \text{ m}$$

The time needed to travel 800×10^6 km $= 8.0 \times 10^{11}$ m is

$$\Delta t = \frac{\Delta x}{c} = \frac{8.0 \times 10^{11} \text{ m}}{3.00 \times 10^8 \text{ m/s}} = 2700 \text{ s} = 45 \text{ min}$$

The Index of Refraction

Light waves travel with speed c in a vacuum, but they slow down as they pass through transparent materials such as water or glass or even, to a very slight extent, air. The slowdown is a consequence of interactions between the electromagnetic field of the wave and the electrons in the material. The speed of light in a material is characterized by the material's **index of refraction** n, defined as

$$n = \frac{\text{speed of light in a vacuum}}{\text{speed of light in the material}} = \frac{c}{v} \tag{20.29}$$

The index of refraction of a material is always greater than 1 because $v < c$. A vacuum has $n = 1$ exactly. Table 20.2 shows the index of refraction for several materials. You can see that liquids and solids have larger indices of refraction than gases.

NOTE ▶ An accurate value for the index of refraction of air is relevant only in very precise measurements. We will assume $n_{air} = 1.00$ in this text. ◀

If the speed of a light wave changes as it enters into a transparent material, such as glass, what happens to the light's frequency and wavelength? Because $v = \lambda f$, either λ or f or both have to change when v changes.

As an analogy, think of a sound wave in the air as it impinges on the surface of a pool of water. As the air oscillates back and forth, it periodically pushes on the surface of the water. These pushes generate the compressions of the sound wave that continues on into the water. Because each push of the air causes one compression of the water, the frequency of the sound wave in the water must be *exactly the same* as the frequency of the sound wave in the air. In other words, **the frequency of a wave is the frequency of the source. It does not change as the wave moves from one medium to another.**

The same is true for electromagnetic waves; the frequency does not change as the wave moves from one material to another.

FIGURE 20.24 shows a light wave passing through a transparent material with index of refraction n. As the wave travels through vacuum it has wavelength λ_{vac} and frequency f_{vac} such that $\lambda_{vac} f_{vac} = c$. In the material, $\lambda_{mat} f_{mat} = v = c/n$. The frequency does not change as the wave enters ($f_{mat} = f_{vac}$), so the wavelength must. The wavelength in the material is

$$\lambda_{mat} = \frac{v}{f_{mat}} = \frac{c}{n f_{mat}} = \frac{c}{n f_{vac}} = \frac{\lambda_{vac}}{n} \qquad (20.30)$$

The wavelength in the transparent material is less than the wavelength in vacuum. This makes sense. Suppose a marching band is marching at one step per second at a speed of 1 m/s. Suddenly they slow their speed to $\frac{1}{2}$ m/s but maintain their march at one step per second. The only way to go slower while marching at the same pace is to take *smaller steps*. When a light wave enters a material, the only way it can go slower while oscillating at the same frequency is to have a *smaller wavelength*.

TABLE 20.2 Typical indices of refraction

Material	Index of refraction
Vacuum	1 exactly
Air	1.0003
Water	1.33
Glass	1.50
Diamond	2.42

FIGURE 20.24 Light passing through a transparent material with index of refraction n.

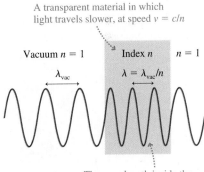

A transparent material in which light travels slower, at speed $v = c/n$

Vacuum $n = 1$ Index n $n = 1$

λ_{vac} $\lambda = \lambda_{vac}/n$

The wavelength inside the material decreases, but the frequency doesn't change.

EXAMPLE 20.8 | **Light traveling through glass**

Orange light with a wavelength of 600 nm is incident upon a 1.00-mm-thick glass microscope slide.

a. What is the light speed in the glass?
b. How many wavelengths of the light are inside the slide?

SOLVE a. From Table 20.2 we see that the index of refraction of glass is $n_{glass} = 1.50$. Thus the speed of light in glass is

$$v_{glass} = \frac{c}{n_{glass}} = \frac{3.00 \times 10^8 \text{ m/s}}{1.50} = 2.00 \times 10^8 \text{ m/s}$$

b. The wavelength inside the glass is

$$\lambda_{glass} = \frac{\lambda_{vac}}{n_{glass}} = \frac{600 \text{ nm}}{1.50} = 400 \text{ nm} = 4.00 \times 10^{-7} \text{ m}$$

N wavelengths span a distance $d = N\lambda$, so the number of wavelengths in $d = 1.00$ mm is

$$N = \frac{d}{\lambda} = \frac{1.00 \times 10^{-3} \text{ m}}{4.00 \times 10^{-7} \text{ m}} = 2500$$

ASSESS The fact that 2500 wavelengths fit within 1 mm shows how small the wavelengths of light are.

STOP TO THINK 20.5 A light wave travels through three transparent materials of equal thickness. Rank in order, from largest to smallest, the indices of refraction n_a, n_b, and n_c.

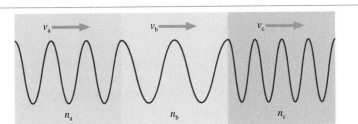

20.6 Power, Intensity, and Decibels

A traveling wave transfers energy from one point to another. The sound wave from a loudspeaker sets your eardrum into motion. Light waves from the sun warm the earth. The *power* of a wave is the rate, in joules per second, at which the wave transfers energy. As you learned in Chapter 11, power is measured in watts. A loudspeaker might emit 2 W of power, meaning that energy in the form of sound waves is radiated at the rate of 2 joules per second. A lightbulb might emit 5 W, or 5 J/s, of visible light. (In fact, this is about right for a so-called 100 watt bulb, with the other 95 W of power being emitted as heat, or infrared radiation, rather than as visible light.)

Imagine doing two experiments with a lightbulb that emits 5 W of visible light. In the first, you hang the bulb in the center of a room and allow the light to illuminate the walls. In the second experiment, you use mirrors and lenses to "capture" the bulb's light and focus it onto a small spot on one wall. This is what a computer projector does. The energy emitted by the bulb is the same in both cases, but, as you know, the light is much brighter when focused onto a small area. We would say that the focused light is more *intense* than the diffuse light that goes in all directions. Similarly, a loudspeaker that beams its sound forward into a small area produces a louder sound in that area than a speaker of equal power that radiates the sound in all directions. Quantities such as brightness and loudness depend not only on the rate of energy transfer, or power, but also on the *area* that receives that power.

FIGURE 20.25 shows a wave impinging on a surface of area a. The surface is perpendicular to the direction in which the wave is traveling. This might be a real, physical surface, such as your eardrum or a photovoltaic cell, but it could equally well be a mathematical surface in space that the wave passes right through. If the wave has power P, we define the **intensity** I of the wave to be

$$I = \frac{P}{a} = \text{power-to-area ratio} \qquad (20.31)$$

The SI units of intensity are W/m^2. Because intensity is a power-to-area ratio, a wave focused into a small area will have a larger intensity than a wave of equal power that is spread out over a large area.

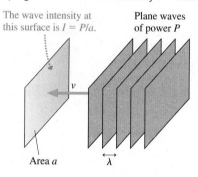

FIGURE 20.25 Plane waves of power P impinge on area a with intensity $I = P/a$.

The wave intensity at this surface is $I = P/a$.

Plane waves of power P

v

Area a

λ

EXAMPLE 20.9 **The intensity of a laser beam**

A helium-neon laser, the kind that provides the familiar red light of classroom demonstrations and supermarket checkout scanners, emits 1.0 mW of light power into a 1.0-mm-diameter laser beam. What is the intensity of the laser beam?

MODEL The laser beam is a light wave.

SOLVE The light waves of the laser beam pass through a mathematical surface that is a circle of diameter 1.0 mm. The intensity of the laser beam is

$$I = \frac{P}{a} = \frac{P}{\pi r^2} = \frac{0.0010 \text{ W}}{\pi (0.00050 \text{ m})^2} = 1300 \text{ W/m}^2$$

ASSESS This is roughly the intensity of sunlight at noon on a summer day. The difference between the sun and a small laser is not their intensities, which are about the same, but their powers. The laser has a small power of 1 mW. It can produce a very intense wave only because the area through which the wave passes is very small. The sun, by contrast, radiates a total power $P_{\text{sun}} \approx 4 \times 10^{26}$ W. This immense power is spread through *all* of space, producing an intensity of 1400 W/m^2 at a distance of 1.5×10^{11} m, the radius of the earth's orbit.

If a source of spherical waves radiates uniformly in all directions, then, as **FIGURE 20.26** shows, the power at distance r is spread uniformly over the surface of a sphere of radius r. The surface area of a sphere is $a = 4\pi r^2$, so the intensity of a uniform spherical wave is

$$I = \frac{P_{\text{source}}}{4\pi r^2} \qquad \text{(intensity of a uniform spherical wave)} \qquad (20.32)$$

The inverse-square dependence of r is really just a statement of energy conservation. The source emits energy at the rate P joules per second. The energy is spread over a larger and larger area as the wave moves outward. Consequently, the energy *per unit area* must decrease in proportion to the surface area of a sphere.

If the intensity at distance r_1 is $I_1 = P_{source}/4\pi r_1^2$ and the intensity at r_2 is $I_2 = P_{source}/4\pi r_2^2$, then you can see that the intensity *ratio* is

$$\frac{I_1}{I_2} = \frac{r_2^2}{r_1^2} \tag{20.33}$$

You can use Equation 20.33 to compare the intensities at two distances from a source without needing to know the power of the source.

> **NOTE ▶** Wave intensities are strongly affected by reflections and absorption. Equations 20.32 and 20.33 apply to situations such as the light from a star or the sound from a firework exploding high in the air. Indoor sound does *not* obey a simple inverse-square law because of the many reflecting surfaces. ◀

For a sinusoidal wave, each particle in the medium oscillates back and forth in simple harmonic motion. You learned in Chapter 14 that a particle in SHM with amplitude A has energy $E = \frac{1}{2}kA^2$, where k is the spring constant of the medium, not the wave number. It is this oscillatory energy of the medium that is transferred, particle to particle, as the wave moves through the medium.

Because a wave's intensity is proportional to the rate at which energy is transferred through the medium, and because the oscillatory energy in the medium is proportional to the *square* of the amplitude, we can infer that

$$I \propto A^2 \tag{20.34}$$

That is, **the intensity of a wave is proportional to the square of its amplitude.** If you double the amplitude of a wave, you increase its intensity by a factor of 4.

Human hearing spans an extremely wide range of intensities, from the *threshold of hearing* at $\approx 1 \times 10^{-12}$ W/m^2 (at midrange frequencies) to the *threshold of pain* at ≈ 10 W/m^2. If we want to make a scale of loudness, it's convenient and logical to place the zero of our scale at the threshold of hearing. To do so, we define the **sound intensity level,** expressed in **decibels** (dB), as

$$\beta = (10 \text{ dB}) \log_{10}\left(\frac{I}{I_0}\right) \tag{20.35}$$

where $I_0 = 1.0 \times 10^{-12}$ W/m^2. The symbol β is the Greek letter beta. Notice that β is computed as a base-10 logarithm, not a natural logarithm.

The decibel is named after Alexander Graham Bell, inventor of the telephone. Sound intensity level is actually dimensionless because it's formed from the ratio of two intensities, so decibels are just a *name* to remind us that we're dealing with an intensity *level* rather than a true intensity.

Right at the threshold of hearing, where $I = I_0$, the sound intensity level is

$$\beta = (10 \text{ dB}) \log_{10}\left(\frac{I_0}{I_0}\right) = (10 \text{ dB}) \log_{10}(1) = 0 \text{ dB}$$

Note that 0 dB doesn't mean no sound; it means that, for most people, no sound is heard. Dogs have more sensitive hearing than humans, and most dogs can easily perceive a 0 dB sound. The sound intensity level at the pain threshold is

$$\beta = (10 \text{ dB}) \log_{10}\left(\frac{10 \text{ W/m}^2}{10^{-12} \text{ W/m}^2}\right) = (10 \text{ dB}) \log_{10}(10^{13}) = 130 \text{ dB}$$

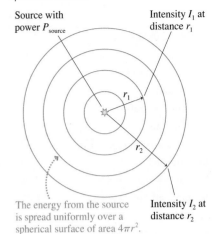

FIGURE 20.26 A source emitting uniform spherical waves.

Source with power P_{source}

Intensity I_1 at distance r_1

r_1

r_2

The energy from the source is spread uniformly over a spherical surface of area $4\pi r^2$.

Intensity I_2 at distance r_2

TABLE 20.3 Sound intensity levels of common sounds

Sound	β (dB)
Threshold of hearing	0
Person breathing, at 3 m	10
A whisper, at 1 m	20
Quiet room	30
Outdoors, no traffic	40
Quiet restaurant	50
Normal conversation, at 1 m	60
Busy traffic	70
Vacuum cleaner, for user	80
Niagara Falls, at viewpoint	90
Snowblower, at 2 m	100
Stereo, at maximum volume	110
Rock concert	120
Threshold of pain	130

The major point to notice is that the sound intensity level increases by 10 dB each time the actual intensity increases by a *factor* of 10. For example, the sound intensity level increases from 70 dB to 80 dB when the sound intensity increases from 10^{-5} W/m² to 10^{-4} W/m². Perception experiments find that sound is perceived as "twice as loud" when the intensity increases by a factor of 10. In terms of decibels, we can say that the perceived loudness of a sound doubles with each increase in the sound intensity level by 10 dB.

Table 20.3 gives the sound intensity levels for a number of sounds. Although 130 dB is the threshold of pain, quieter sounds can damage your hearing. A fairly short exposure to 120 dB can cause damage to the hair cells in the ear, but lengthy exposure to sound intensity levels of over 85 dB can produce damage as well.

EXAMPLE 20.10 **Blender noise**

The blender making a smoothie produces a sound intensity level of 83 dB. What is the intensity of the sound? What will the sound intensity level be if a second blender is turned on?

SOLVE We can solve Equation 20.35 for the sound intensity, finding $I = I_0 \times 10^{\beta/10 \text{ dB}}$. Here we used the fact that 10 raised to a power is an "antilogarithm." In this case,

$$I = (1.0 \times 10^{-12} \text{ W/m}^2) \times 10^{8.3} = 2.0 \times 10^{-4} \text{ W/m}^2$$

A second blender doubles the sound power and thus raises the intensity to $I = 4.0 \times 10^{-4}$ W/m². The new sound intensity level is

$$\beta = (10 \text{ dB}) \log_{10}\left(\frac{4.0 \times 10^{-4} \text{ W/m}^2}{1.0 \times 10^{-12} \text{ W/m}^2}\right) = 86 \text{ dB}$$

ASSESS In general, doubling the actual sound intensity increases the decibel level by 3 dB.

STOP TO THINK 20.6 Four trumpet players are playing the same note. If three of them suddenly stop, the sound intensity level decreases by

a. 40 dB   b. 12 dB   c. 6 dB   d. 4 dB

20.7 The Doppler Effect

Our final topic for this chapter is an interesting effect that occurs when you are in motion relative to a wave source. It is called the *Doppler effect.* You've likely noticed that the pitch of an ambulance's siren drops as it goes past you. Why?

FIGURE 20.27a shows a source of sound waves moving away from Pablo and toward Nancy at a steady speed v_s. The subscript s indicates that this is the speed of the source, not the speed of the waves. The source is emitting sound waves of frequency f_0 as it travels. The figure is a motion diagram showing the position of the source at times $t = 0$, T, $2T$, and $3T$, where $T = 1/f_0$ is the period of the waves.

Nancy measures the frequency of the wave emitted by the *approaching source* to be f_+. At the same time, Pablo measures the frequency of the wave emitted by the *receding source* to be f_-. Our task is to relate f_+ and f_- to the source frequency f_0 and speed v_s.

After a wave crest leaves the source, its motion is governed by the properties of the medium. That is, the motion of the source cannot affect a wave that has already been emitted. Thus each circular wave front in FIGURE 20.27b is centered on the point from which it was emitted. The wave crest from point 3 was emitted just as this figure was made, but it hasn't yet had time to travel any distance.

The wave crests are bunched up in the direction the source is moving, stretched out behind it. The distance between one crest and the next is one wavelength, so the wavelength λ_+ Nancy measures is *less* than the wavelength $\lambda_0 = v/f_0$ that would be emitted if the source were at rest. Similarly, λ_- behind the source is larger than λ_0.

FIGURE 20.27 A motion diagram showing the wave fronts emitted by a source as it moves to the right at speed v_s.

(a) Motion of the source

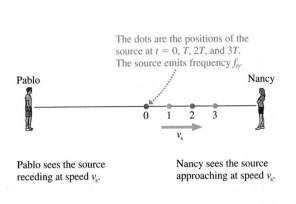

The dots are the positions of the source at $t = 0$, T, $2T$, and $3T$. The source emits frequency f_0.

Pablo

Nancy

v_s

Pablo sees the source receding at speed v_s.

Nancy sees the source approaching at speed v_s.

(b) Snapshot at time $3T$

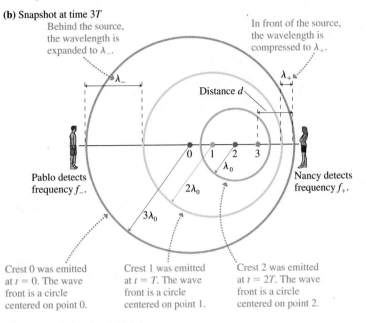

Behind the source, the wavelength is expanded to λ_-.

In front of the source, the wavelength is compressed to λ_+.

λ_-

λ_+

Distance d

λ_0

$2\lambda_0$

$3\lambda_0$

Pablo detects frequency f_-.

Nancy detects frequency f_+.

Crest 0 was emitted at $t = 0$. The wave front is a circle centered on point 0.

Crest 1 was emitted at $t = T$. The wave front is a circle centered on point 1.

Crest 2 was emitted at $t = 2T$. The wave front is a circle centered on point 2.

These crests move through the medium at the wave speed v. Consequently, the frequency $f_+ = v/\lambda_+$ detected by the observer whom the source is approaching is *higher* than the frequency f_0 emitted by the source. Similarly, $f_- = v/\lambda_-$ detected behind the source is *lower* than frequency f_0. This change of frequency when a source moves relative to an observer is called the **Doppler effect.**

The distance labeled d in Figure 20.27b is the difference between how far the wave has moved and how far the source has moved at time $t = 3T$. These distances are

$$\Delta x_{\text{wave}} = vt = 3vT$$
$$\Delta x_{\text{source}} = v_s t = 3v_s T \tag{20.36}$$

The distance d spans three wavelengths; thus the wavelength of the wave emitted by an approaching source is

$$\lambda_+ = \frac{d}{3} = \frac{\Delta x_{\text{wave}} - \Delta x_{\text{source}}}{3} = \frac{3vT - 3v_s T}{3} = (v - v_s)T \tag{20.37}$$

You can see that our arbitrary choice of three periods was not relevant because the 3 cancels. The frequency detected in Nancy's direction is

$$f_+ = \frac{v}{\lambda_+} = \frac{v}{(v - v_s)T} = \frac{v}{(v - v_s)}f_0 \tag{20.38}$$

where $f_0 = 1/T$ is the frequency of the source and is the frequency you would detect if the source were at rest. We'll find it convenient to write the detected frequency as

$$f_+ = \frac{f_0}{1 - v_s/v} \quad \text{(Doppler effect for an approaching source)}$$
$$\tag{20.39}$$
$$f_- = \frac{f_0}{1 + v_s/v} \quad \text{(Doppler effect for a receding source)}$$

Proof of the second version, for the frequency f_- of a receding source, is similar. You can see that $f_+ > f_0$ in front of the source, because the denominator is less than 1, and $f_- < f_0$ behind the source.

Doppler weather radar uses the Doppler shift of reflected radar signals to measure wind speeds and thus better gauge the severity of a storm.

EXAMPLE 20.11 **How fast are the police traveling?**

A police siren has a frequency of 550 Hz as the police car approaches you, 450 Hz after it has passed you and is receding. How fast are the police traveling? The temperature is 20°C.

MODEL The siren's frequency is altered by the Doppler effect. The frequency is f_+ as the car approaches and f_- as it moves away.

SOLVE To find v_s, we rewrite Equations 20.39 as

$$f_0 = (1 + v_s/v)f_-$$

$$f_0 = (1 - v_s/v)f_+$$

We subtract the second equation from the first, giving

$$0 = f_- - f_+ + \frac{v_s}{v}(f_- + f_+)$$

This is easily solved to give

$$v_s = \frac{f_+ - f_-}{f_+ + f_-}v = \frac{100 \text{ Hz}}{1000 \text{ Hz}} 343 \text{ m/s} = 34.3 \text{ m/s}$$

ASSESS If you now solve for the siren frequency when at rest, you will find $f_0 = 495$ Hz. Surprisingly, the at-rest frequency is not halfway between f_- and f_+.

A Stationary Source and a Moving Observer

Suppose the police car in Example 20.11 is at rest while you drive toward it at 34.3 m/s. You might think that this is equivalent to having the police car move toward you at 34.3 m/s, but it isn't. Mechanical waves move through a medium, and the Doppler effect depends not just on how the source and the observer move with respect to each other but also on how they move with respect to the medium. We'll omit the proof, but it's not hard to show that the frequencies heard by an observer moving at speed v_o relative to a stationary source emitting frequency f_0 are

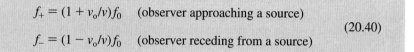

$$f_+ = (1 + v_o/v)f_0 \quad \text{(observer approaching a source)}$$
$$f_- = (1 - v_o/v)f_0 \quad \text{(observer receding from a source)} \tag{20.40}$$

A quick calculation shows that the frequency of the police siren as you approach it at 34.3 m/s is 545 Hz, not the 550 Hz you heard as it approached you at 34.3 m/s.

The Doppler Effect for Light Waves

The Doppler effect is observed for all types of waves, not just sound waves. If a source of light waves is receding from you, the wavelength λ_- that you detect is longer than the wavelength λ_0 emitted by the source.

Although the reason for the Doppler shift for light is the same as for sound waves, there is one fundamental difference. We derived Equations 20.39 for the Doppler-shifted frequencies by measuring the wave speed v relative to the medium. For electromagnetic waves in empty space, there is no medium. Consequently, we need to turn to Einstein's theory of relativity to determine the frequency of light waves from a moving source. The result, which we state without proof, is

$$\lambda_- = \sqrt{\frac{1 + v_s/c}{1 - v_s/c}}\,\lambda_0 \quad \text{(receding source)}$$

$$\lambda_+ = \sqrt{\frac{1 - v_s/c}{1 + v_s/c}}\,\lambda_0 \quad \text{(approaching source)} \tag{20.41}$$

Here v_s is the speed of the source *relative to* the observer.

The light waves from a receding source are shifted to longer wavelengths ($\lambda_- > \lambda_0$). Because the longest visible wavelengths are perceived as the color red, the light from a receding source is **red shifted.** That is *not* to say that the light is red, simply that its wavelength is shifted toward the red end of the spectrum. If $\lambda_0 = 470$ nm (blue) light emitted by a rapidly receding source is detected at $\lambda_- = 520$ nm (green), we would say that the light has been red shifted. Similarly, light from an approaching source is **blue shifted,** meaning that the detected wavelengths are shorter than the emitted wavelengths ($\lambda_+ < \lambda_0$) and thus are shifted toward the blue end of the spectrum.

EXAMPLE 20.12 **Measuring the velocity of a galaxy**

Hydrogen atoms in the laboratory emit red light with wavelength 656 nm. In the light from a distant galaxy, this "spectral line" is observed at 691 nm. What is the speed of this galaxy relative to the earth?

MODEL The observed wavelength is longer than the wavelength emitted by atoms at rest with respect to the observer (i.e., red shifted), so we are looking at light emitted from a galaxy that is receding from us.

SOLVE Squaring the expression for λ_- in Equations 20.41 and solving for v_s give

$$v_s = \frac{(\lambda_-/\lambda_0)^2 - 1}{(\lambda_-/\lambda_0)^2 + 1} c$$

$$= \frac{(691 \text{ nm}/656 \text{ nm})^2 - 1}{(691 \text{ nm}/656 \text{ nm})^2 + 1} c$$

$$= 0.052c = 1.56 \times 10^7 \text{ m/s}$$

ASSESS The galaxy is moving away from the earth at about 5% of the speed of light!

In the 1920s, an analysis of the red shifts of many galaxies led the astronomer Edwin Hubble to the conclusion that the galaxies of the universe are *all* moving apart from each other. Extrapolating backward in time must bring us to a point when all the matter of the universe—and even space itself, according to the theory of relativity—began rushing out of a primordial fireball. Many observations and measurements since have given support to the idea that the universe began in a *Big Bang* about 14 billion years ago.

As an example, **FIGURE 20.28** is a Hubble Space Telescope picture of a *quasar*, short for *quasistellar object*. Quasars are extraordinarily powerful sources of light and radio waves. The light reaching us from quasars is highly red shifted, corresponding in some cases to objects that are moving away from us at greater than 90% of the speed of light. Astronomers have determined that some quasars are 10 to 12 *billion* light years away from the earth, hence the light we see was emitted when the universe was only about 25% of its present age. Today, the red shifts of distant quasars and supernovae (exploding stars) are being used to refine our understanding of the structure and evolution of the universe.

FIGURE 20.28 A Hubble Space Telescope picture of a quasar.

STOP TO THINK 20.7 Amy and Zack are both listening to the source of sound waves that is moving to the right. Compare the frequencies each hears.

a. $f_{\text{Amy}} > f_{\text{Zack}}$
b. $f_{\text{Amy}} = f_{\text{Zack}}$
c. $f_{\text{Amy}} < f_{\text{Zack}}$

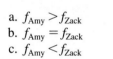

Amy   10 m/s f_0 10 m/s   10 m/s   Zack

Decreasing the sound

The loudspeaker on a homecoming float—mounted on a pole—is stuck playing an annoying 210 Hz tone. When the speaker is 10 m away, you measure the sound to be a loud 95 dB at 208 Hz. How long will it take for the sound intensity level to drop to a tolerable 55 dB?

MODEL The source is on a pole, so model the sound waves as uniform spherical waves. Assume a temperature of 20°C.

SOLVE The 208 Hz frequency you measure is less than the 210 Hz frequency that was emitted, so the float must be moving away from you. The Doppler effect for a receding source is

$$f_- = \frac{f_0}{1 + v_s/v}$$

We can solve this to find the speed of the float:

$$v_s = \left(\frac{f_0}{f_-} - 1\right)v = \left(\frac{210 \text{ Hz}}{208 \text{ Hz}} - 1\right) \times 343 \text{ m/s} = 3.3 \text{ m/s}$$

The sound intensity of a spherical wave decreases with the inverse square of the distance from the source. A sound intensity level β corresponds to an intensity $I = I_0 \times 10^{\beta/10 \text{ dB}}$, where $I_0 = 1.0 \times 10^{-12}$ W/m^2. At the initial 95 dB, the intensity is

$$I_1 = I_0 \times 10^{9.5} = 3.2 \times 10^{-3} \text{ W/m}^2$$

At the desired 55 dB, the intensity will have dropped to

$$I_2 = I_0 \times 10^{5.5} = 3.2 \times 10^{-7} \text{ W/m}^2$$

The intensity ratio is related to the distances by

$$\frac{I_1}{I_2} = \frac{r_2^2}{r_1^2}$$

Thus the sound will have dropped to 55 dB when the distance to the speaker is

$$r_2 = \sqrt{\frac{I_1}{I_2}} r_1 = \sqrt{10^4} \times 10 \text{ m} = 1000 \text{ m}$$

The float has to travel $\Delta x = 990$ m, which will take

$$\Delta t = \frac{\Delta x}{v_s} = \frac{990 \text{ m}}{3.3 \text{ m/s}} = 300 \text{ s} = 5.0 \text{ min}$$

ASSESS To drop the sound intensity level by 40 dB requires decreasing the intensity by a factor of 10^4. And with the intensity depending on the inverse square of the distance, that requires increasing the distance by a factor of 100. Floats don't move very fast—3.3 m/s is about 7 mph—so needing several minutes to travel the ≈ 1000 m seems reasonable.

SUMMARY

The goal of Chapter 20 has been to learn the basic properties of traveling waves.

General Principles

The Wave Model

This model is based on the idea of a traveling wave, which is an organized disturbance traveling at a well-defined **wave speed** v.

- In transverse waves the displacement is perpendicular to the direction in which the wave travels.

- In longitudinal waves the particles of the medium move parallel to the direction in which the wave travels.

A wave transfers **energy,** but no material or substance is transferred outward from the source.

Two basic types of waves:

- Mechanical waves travel through a material medium such as water or air.

- Electromagnetic waves require no material medium and can travel through a vacuum.

For mechanical waves, such as sound waves and waves on strings, the speed of the wave is a property of the medium. Speed does not depend on the size or shape of the wave.

Important Concepts

The **displacement** D of a wave is a function of both position (where) and time (when).

- A snapshot graph shows the wave's displacement as a function of position at a single instant of time.

- A history graph shows the wave's displacement as a function of time at a single point in space.

For a transverse wave on a string, the snapshot graph is a picture of the wave. The displacement of a longitudinal wave is parallel to the motion; thus the snapshot graph of a longitudinal sound wave is *not* a picture of the wave.

Sinusoidal waves are periodic in both time (period T) and space (wavelength λ):

$$D(x, t) = A \sin\left[2\pi(x/\lambda - t/T) + \phi_0\right]$$
$$= A \sin(kx - \omega t + \phi_0)$$

where A is the **amplitude,** $k = 2\pi/\lambda$ is the **wave number,** $\omega = 2\pi f = 2\pi/T$ is the **angular frequency,** and ϕ_0 is the **phase constant** that describes initial conditions.

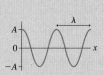

One-dimensional waves

Two- and three-dimensional waves

The fundamental relationship for any sinusoidal wave is $v = \lambda f$.

Applications

- **String** (transverse): $v = \sqrt{T_s/\mu}$
- **Sound** (longitudinal): $v = 343$ m/s in 20°C air
- **Light** (transverse): $v = c/n$, where $c = 3.00 \times 10^8$ m/s is the speed of light in a vacuum and n is the material's **index of refraction**

The wave intensity is the power-to-area ratio: $I = P/a$

For a circular or spherical wave: $I = P_{\text{source}}/4\pi r^2$

The sound intensity level is

$$\beta = (10 \text{ dB}) \log_{10}(I/1.0 \times 10^{-12} \text{ W/m}^2)$$

The Doppler effect occurs when a wave source and detector are moving with respect to each other: the frequency detected differs from the frequency f_0 emitted.

Approaching source

$$f_+ = \frac{f_0}{1 - v_s/v}$$

Receding source

$$f_- = \frac{f_0}{1 + v_s/v}$$

Observer approaching a source

$$f_+ = (1 + v_o/v)f_0$$

Observer receding from a source

$$f_- = (1 - v_o/v)f_0$$

The Doppler effect for light uses a result derived from the theory of relativity.

Terms and Notation

wave model	wave speed, v	wave front	index of refraction, n
traveling wave	linear density, μ	circular wave	intensity, I
transverse wave	snapshot graph	spherical wave	sound intensity level, β
longitudinal wave	history graph	plane wave	decibels
mechanical waves	sinusoidal wave	phase, ϕ	Doppler effect
electromagnetic waves	amplitude, A	compression	red shifted
medium	wavelength, λ	rarefaction	blue shifted
disturbance	wave number, k	electromagnetic spectrum	

CONCEPTUAL QUESTIONS

1. The three wave pulses in **FIGURE Q20.1** travel along the same stretched string. Rank in order, from largest to smallest, their wave speeds v_a, v_b, and v_c. Explain.

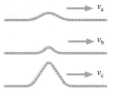

FIGURE Q20.1

2. A wave pulse travels along a stretched string at a speed of 200 cm/s. What will be the speed if:
   a. The string's tension is doubled?
   b. The string's mass is quadrupled (but its length is unchanged)?
   c. The string's length is quadrupled (but its mass is unchanged)?
   Note: Each part is independent and refers to changes made to the original string.

3. **FIGURE Q20.3** is a history graph showing the displacement as a function of time at one point on a string. Did the displacement at this point reach its maximum of 2 mm *before* or *after* the interval of time when the displacement was a constant 1 mm?

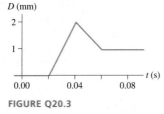

FIGURE Q20.3

4. **FIGURE Q20.4** shows a snapshot graph *and* a history graph for a wave pulse on a stretched string. They describe the same wave from two perspectives.
   a. In which direction is the wave traveling? Explain.
   b. What is the speed of this wave?

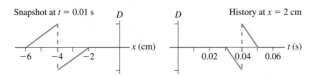

FIGURE Q20.4

5. Rank in order, from largest to smallest, the wavelengths λ_a, λ_b, and λ_c for sound waves having frequencies $f_a = 100$ Hz, $f_b = 1000$ Hz, and $f_c = 10,000$ Hz. Explain.

6. A sound wave with wavelength λ_0 and frequency f_0 moves into a new medium in which the speed of sound is $v_1 = 2v_0$. What are the new wavelength λ_1 and frequency f_1? Explain.

7. What are the amplitude, wavelength, frequency, and phase constant of the traveling wave in **FIGURE Q20.7**?

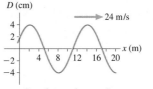

FIGURE Q20.7

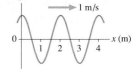

FIGURE Q20.8

8. **FIGURE Q20.8** is a snapshot graph of a sinusoidal wave at $t = 1.0$ s. What is the phase constant of this wave?

9. **FIGURE Q20.9** shows the wave fronts of a circular wave. What is the phase difference between (a) points A and B, (b) points C and D, and (c) points E and F?

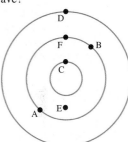

FIGURE Q20.9

10. Sound wave A delivers 2 J of energy in 2 s. Sound wave B delivers 10 J of energy in 5 s. Sound wave C delivers 2 mJ of energy in 1 ms. Rank in order, from largest to smallest, the sound powers P_A, P_B, and P_C of these three sound waves. Explain.

11. One physics professor talking produces a sound intensity level of 52 dB. It's a frightening idea, but what would be the sound intensity level of 100 physics professors talking simultaneously?

12. You are standing at $x = 0$ m, listening to a sound that is emitted at frequency f_0. The graph of **FIGURE Q20.12** shows the frequency you hear during a 4-second interval. Which of the following describes the sound source? Explain your choice.

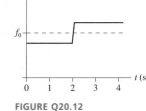

FIGURE Q20.12

   A. It moves from left to right and passes you at $t = 2$ s.
   B. It moves from right to left and passes you at $t = 2$ s.
   C. It moves toward you but doesn't reach you. It then reverses direction at $t = 2$ s.
   D. It moves away from you until $t = 2$ s. It then reverses direction and moves toward you but doesn't reach you.

EXERCISES AND PROBLEMS

Problems labeled ▨ integrate material from earlier chapters.

Exercises

Section 20.1 The Wave Model

1. | The wave speed on a string is 150 m/s when the tension is 75 N. What tension will give a speed of 180 m/s?

2. | The wave speed on a string under tension is 200 m/s. What is the speed if the tension is halved?

3. ‖ A 25 g string is under 20 N of tension. A pulse travels the length of the string in 50 ms. How long is the string?

Section 20.2 One-Dimensional Waves

4. ‖ Draw the history graph $D(x = 0$ m, $t)$ at $x = 0$ m for the wave shown in FIGURE EX20.4.

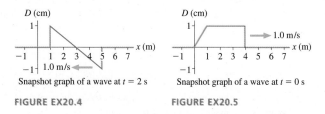

Snapshot graph of a wave at $t = 2$ s

FIGURE EX20.4

Snapshot graph of a wave at $t = 0$ s

FIGURE EX20.5

5. ‖ Draw the history graph $D(x = 5.0$ m, $t)$ at $x = 5.0$ m for the wave shown in FIGURE EX20.5.

6. ‖ Draw the snapshot graph $D(x, t = 0$ s$)$ at $t = 0$ s for the wave shown in FIGURE EX20.6.

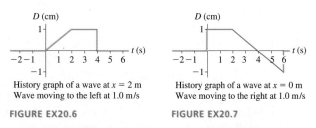

History graph of a wave at $x = 2$ m
Wave moving to the left at 1.0 m/s

FIGURE EX20.6

History graph of a wave at $x = 0$ m
Wave moving to the right at 1.0 m/s

FIGURE EX20.7

7. ‖ Draw the snapshot graph $D(x, t = 1.0$ s$)$ at $t = 1.0$ s for the wave shown in FIGURE EX20.7.

8. ‖ FIGURE EX20.8 is the snapshot graph at $t = 0$ s of a *longitudinal* wave. Draw the corresponding picture of the particle positions, as was done in Figure 20.10b. Let the equilibrium spacing between the particles be 1.0 cm.

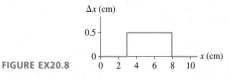

FIGURE EX20.8

9. ‖ FIGURE EX20.9 is a picture at $t = 0$ s of the particles in a medium as a longitudinal wave is passing through. The equilibrium spacing between the particles is 1.0 cm. Draw the snapshot graph $D(x, t = 0$ s$)$ of this wave at $t = 0$ s.

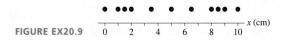

FIGURE EX20.9

Section 20.3 Sinusoidal Waves

10. | A wave travels with speed 200 m/s. Its wave number is 1.5 rad/m. What are its (a) wavelength and (b) frequency?

11. | A wave has angular frequency 30 rad/s and wavelength 2.0 m. What are its (a) wave number and (b) wave speed?

12. | The displacement of a wave traveling in the positive x-direction is $D(x, t) = (3.5$ cm$)\sin(2.7x - 124t)$, where x is in m and t is in s. What are the (a) frequency, (b) wavelength, and (c) speed of this wave?

13. | The displacement of a wave traveling in the negative y-direction is $D(y, t) = (5.2$ cm$)\sin(5.5y + 72t)$, where y is in m and t is in s. What are the (a) frequency, (b) wavelength, and (c) speed of this wave?

14. | What are the amplitude, frequency, and wavelength of the wave in FIGURE EX20.14?

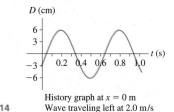

FIGURE EX20.14

History graph at $x = 0$ m
Wave traveling left at 2.0 m/s

Section 20.4 Waves in Two and Three Dimensions

15. | A spherical wave with a wavelength of 2.0 m is emitted from the origin. At one instant of time, the phase at $r = 4.0$ m is π rad. At that instant, what is the phase at $r = 3.5$ m and at $r = 4.5$ m?

16. | A circular wave travels outward from the origin. At one instant of time, the phase at $r_1 = 20$ cm is 0 rad and the phase at $r_2 = 80$ cm is 3π rad. What is the wavelength of the wave?

17. ‖ A loudspeaker at the origin emits a 120 Hz tone on a day when the speed of sound is 340 m/s. The phase difference between two points on the x-axis is 5.5 rad. What is the distance between these two points?

18. ‖ A sound source is located somewhere along the x-axis. Experiments show that the same wave front simultaneously reaches listeners at $x = -7.0$ m and $x = +3.0$ m.
   a. What is the x-coordinate of the source?
   b. A third listener is positioned along the positive y-axis. What is her y-coordinate if the same wave front reaches her at the same instant it does the first two listeners?

Section 20.5 Sound and Light

19. ‖ A hammer taps on the end of a 4.00-m-long metal bar at room temperature. A microphone at the other end of the bar picks up two pulses of sound, one that travels through the metal and one that travels through the air. The pulses are separated in time by 9.00 ms. What is the speed of sound in this metal?

20. ‖ a. What is the wavelength of a 2.0 MHz ultrasound wave traveling through aluminum?
   b. What frequency of electromagnetic wave would have the same wavelength as the ultrasound wave of part a?

21. | a. What is the frequency of an electromagnetic wave with a wavelength of 20 cm?
    b. What would be the wavelength of a sound wave in water with the same frequency as the electromagnetic wave of part a?

22. | a. What is the frequency of blue light that has a wavelength of 450 nm?
    b. What is the frequency of red light that has a wavelength of 650 nm?
    c. What is the index of refraction of a material in which the red-light wavelength is 450 nm?

23. | a. An FM radio station broadcasts at a frequency of 101.3 MHz. What is the wavelength?
    b. What is the frequency of a sound source that produces the same wavelength in 20°C air?

24. | a. Telephone signals are often transmitted over long distances by microwaves. What is the frequency of microwave radiation with a wavelength of 3.0 cm?
    b. Microwave signals are beamed between two mountaintops 50 km apart. How long does it take a signal to travel from one mountaintop to the other?

25. || a. How long does it take light to travel through a 3.0-mm-thick piece of window glass?
    b. Through what thickness of water could light travel in the same amount of time?

26. || Cell phone conversations are transmitted by high-frequency radio waves. Suppose the signal has wavelength 35 cm while traveling through air. What are the (a) frequency and (b) wavelength as the signal travels through 3-mm-thick window glass into your room?

27. | A light wave has a 670 nm wavelength in air. Its wavelength in a transparent solid is 420 nm.
    a. What is the speed of light in this solid?
    b. What is the light's frequency in the solid?

Section 20.6 Power, Intensity, and Decibels

28. || A sound wave with intensity 2.0×10^{-3} W/m^2 is perceived to
BIO be modestly loud. Your eardrum is 6.0 mm in diameter. How much energy will be transferred to your eardrum while listening to this sound for 1.0 min?

29. || The intensity of electromagnetic waves from the sun is 1.4 kW/m^2 just above the earth's atmosphere. Eighty percent of this reaches the surface at noon on a clear summer day. Suppose you think of your back as a 30 cm $\times$ 50 cm rectangle. How many joules of solar energy fall on your back as you work on your tan for 1.0 h?

30. || A concert loudspeaker suspended high above the ground emits 35 W of sound power. A small microphone with a 1.0 cm^2 area is 50 m from the speaker.
    a. What is the sound intensity at the position of the microphone?
    b. How much sound energy impinges on the microphone each second?

31. || During takeoff, the sound intensity level of a jet engine is 140 dB at a distance of 30 m. What is the sound intensity level at a distance of 1.0 km?

32. | The sun emits electromagnetic waves with a power of 4.0×10^{26} W. Determine the intensity of electromagnetic waves from the sun just outside the atmospheres of Venus, the earth, and Mars.

33. | What are the sound intensity levels for sound waves of intensity (a) 3.0×10^{-6} W/m^2 and (b) 3.0×10^{-2} W/m^2?

34. | What are the intensities of sound waves with sound intensity levels (a) 46 dB and (b) 103 dB?

35. || A loudspeaker on a tall pole broadcasts sound waves equally in all directions. What is the speaker's power output if the sound intensity level is 90 dB at a distance of 20 m?

Section 20.7 The Doppler Effect

36. | A friend of yours is loudly singing a single note at 400 Hz while racing toward you at 25.0 m/s on a day when the speed of sound is 340 m/s.
    a. What frequency do you hear?
    b. What frequency does your friend hear if you suddenly start singing at 400 Hz?

37. | An opera singer in a convertible sings a note at 600 Hz while cruising down the highway at 90 km/h. What is the frequency heard by
    a. A person standing beside the road in front of the car?
    b. A person on the ground behind the car?

38. || A bat locates insects by emitting ultrasonic "chirps" and then
BIO listening for echoes from the bugs. Suppose a bat chirp has a frequency of 25 kHz. How fast would the bat have to fly, and in what direction, for you to just barely be able to hear the chirp at 20 kHz?

39. | A mother hawk screeches as she dives at you. You recall from biology that female hawks screech at 800 Hz, but you hear the screech at 900 Hz. How fast is the hawk approaching?

Problems

40. || The displacement of a traveling wave is

$$D(x, t) = \begin{cases} 1 \text{ cm} & \text{if } |x - 3t| \leq 1 \\ 0 \text{ cm} & \text{if } |x - 3t| > 1 \end{cases}$$

    where x is in m and t in s.
    a. Draw displacement-versus-position graphs from $x = -2$ m to $x = 12$ m at 1 s intervals from $t = 0$ s to $t = 3$ s.
    b. Determine the wave speed from the graphs. Explain.
    c. Determine the wave speed from the equation for $D(x, t)$. Does it agree with your answer to part b?

41. || **FIGURE P20.41** is a history graph at $x = 0$ m of a wave traveling in the positive x-direction at 4.0 m/s.
    a. What is the wavelength?
    b. What is the phase constant of the wave?
    c. Write the displacement equation for this wave.

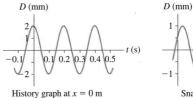

History graph at $x = 0$ m
Wave traveling right at 4.0 m/s

FIGURE P20.41   **FIGURE P20.42**

42. || **FIGURE P20.42** is a snapshot graph at $t = 0$ s of a 5.0 Hz wave traveling to the left.
    a. What is the wave speed?
    b. What is the phase constant of the wave?
    c. Write the displacement equation for this wave.

43. ‖ A wave travels along a string at speed v_0. What will be the speed if the string is replaced by one made of the same material and under the same tension but having twice the radius?

44. | String 1 in FIGURE P20.44 has linear density 2.0 g/m and string 2 has linear density 4.0 g/m. A student sends pulses in both directions by quickly pulling up on the knot, then releasing it. What should the string lengths L_1 and L_2 be if the pulses are to reach the ends of the strings simultaneously?

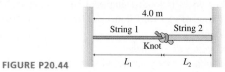

FIGURE P20.44

45. ‖ Ships measure the distance to the ocean bottom with sonar. A pulse of sound waves is aimed at the ocean bottom, then sensitive microphones listen for the echo. FIGURE P20.45 shows the delay time as a function of the ship's position as it crosses 60 km of ocean. Draw a graph of the ocean bottom. Let the ocean surface define $y = 0$ and ocean bottom have negative values of y. This way your graph will be a picture of the ocean bottom. The speed of sound in ocean water varies slightly with temperature, but you can use 1500 m/s as an average value.

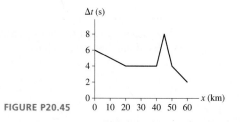

FIGURE P20.45

46. ‖ Oil explorers set off explosives to make loud sounds, then listen for the echoes from underground oil deposits. Geologists suspect that there is oil under 500-m-deep Lake Physics. It's known that Lake Physics is carved out of a granite basin. Explorers detect a weak echo 0.94 s after exploding dynamite at the lake surface. If it's really oil, how deep will they have to drill into the granite to reach it?

47. ‖ One cue your hearing system uses to localize a sound (i.e., to BIO tell where a sound is coming from) is the slight difference in the arrival times of the sound at your ears. Your ears are spaced approximately 20 cm apart. Consider a sound source 5.0 m from the center of your head along a line 45° to your right. What is the difference in arrival times? Give your answer in microseconds. **Hint:** You are looking for the difference between two numbers that are nearly the same. What does this near equality imply about the necessary precision during intermediate stages of the calculation?

48. ‖ A helium-neon laser beam has a wavelength in air of 633 nm. It takes 1.38 ns for the light to travel through 30 cm of an unknown liquid. What is the wavelength of the laser beam in the liquid?

49. | A 440 Hz sound wave in 20°C air propagates into the water of a swimming pool. What are the wave's (a) frequency and (b) wavelength in the water?

50. ‖ Earthquakes are essentially sound waves—called seismic waves—traveling through the earth. Because the earth is solid, it can support both longitudinal and transverse seismic waves. The speed of longitudinal waves, called P waves, is 8000 m/s. Transverse waves, called S waves, travel at a slower 4500 m/s.

A seismograph records the two waves from a distant earthquake. If the S wave arrives 2.0 min after the P wave, how far away was the earthquake? You can assume that the waves travel in straight lines, although actual seismic waves follow more complex routes.

51. ‖ A sound wave is described by $D(y, t) = (0.0200 \text{ mm}) \times \sin\left[(8.96 \text{ rad/m})y + (3140 \text{ rad/s})t + \pi/4 \text{ rad}\right]$, where y is in m and t is in s.
   a. In what direction is this wave traveling?
   b. Along which axis is the air oscillating?
   c. What are the wavelength, the wave speed, and the period of oscillation?

52. ‖ A wave on a string is described by $D(x, t) = (3.0 \text{ cm}) \times \sin\left[2\pi(x/(2.4 \text{ m}) + t/(0.20 \text{ s}) + 1)\right]$, where x is in m and t is in s.
   a. In what direction is this wave traveling?
   b. What are the wave speed, the frequency, and the wave number?
   c. At $t = 0.50$ s, what is the displacement of the string at $x = 0.20$ m?

53. ‖ A wave on a string is described by $D(x, t) = (2.00 \text{ cm}) \times \sin\left[(12.57 \text{ rad/m})x - (638 \text{ rad/s})t\right]$, where x is in m and t in s. The linear density of the string is 5.00 g/m. What are
   a. The string tension?
   b. The maximum displacement of a point on the string?
   c. The maximum speed of a point on the string?

54. | Write the displacement equation for a sinusoidal wave that is traveling in the negative y-direction with wavelength 50 cm, speed 4.0 m/s, and amplitude 5.0 cm. Assume $\phi_0 = 0$.

55. | Write the displacement equation for a sinusoidal wave that is traveling in the positive x-direction with frequency 200 Hz, speed 400 m/s, amplitude 0.010 mm, and phase constant $\pi/2$ rad.

56. | A string with linear density 2.0 g/m is stretched along the positive x-axis with tension 20 N. One end of the string, at $x = 0$ m, is tied to a hook that oscillates up and down at a frequency of 100 Hz with a maximum displacement of 1.0 mm. At $t = 0$ s, the hook is at its lowest point.
   a. What are the wave speed on the string and the wavelength?
   b. What are the amplitude and phase constant of the wave?
   c. Write the equation for the displacement $D(x, t)$ of the traveling wave.
   d. What is the string's displacement at $x = 0.50$ m and $t = 15$ ms?

57. ‖ FIGURE P20.57 shows a snapshot graph of a wave traveling to the right along a string at 45 m/s. At this instant, what is the velocity of points 1, 2, and 3 on the string?

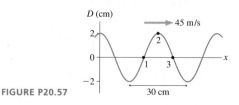

FIGURE P20.57

58. ‖ FIGURE P20.58 shows two masses hanging from a steel wire. The mass of the wire is 60.0 g. A wave pulse travels along the wire from point 1 to point 2 in 24.0 ms. What is mass m?

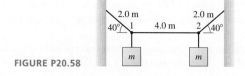

FIGURE P20.58

59. ‖ A wire is made by welding together two metals having different densities. **FIGURE P20.59** shows a 2.00-m-long section of wire centered on the junction, but the wire extends much farther in both directions. The wire is placed under 2250 N tension, then a 1500 Hz wave with an amplitude of 3.00 mm is sent down the wire. How many wavelengths (complete cycles) of the wave are in this 2.00-m-long section of the wire?

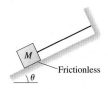

FIGURE P20.59
2250 N ⟵ 1.00 m ⟶⟵ 1.00 m ⟶ 2250 N
$\mu_1 = 9.00$ g/m $\mu_2 = 25.0$ g/m

60. ‖ The string in **FIGURE P20.60** has linear density μ. Find an expression in terms of M, μ and θ for the speed of waves on the string.

FIGURE P20.60
M
Frictionless
θ

61. ‖ A string that is under 50.0 N of tension has linear density 5.0 g/m. A sinusoidal wave with amplitude 3.0 cm and wavelength 2.0 m travels along the string. What is the maximum speed of a particle on the string?

62. ‖ A sinusoidal wave travels along a stretched string. A particle on the string has a maximum speed of 2.0 m/s and a maximum acceleration of 200 m/s². What are the frequency and amplitude of the wave?

63. ‖ a. A 100 W lightbulb produces 5.0 W of visible light. (The other 95 W are dissipated as heat and infrared radiation.) What is the light intensity on a wall 2.0 m away from the lightbulb?
   b. A krypton laser produces a cylindrical red laser beam 2.0 mm in diameter with 5.0 W of power. What is the light intensity on a wall 2.0 m away from the laser?

64. ‖ An AM radio station broadcasts with a power of 25 kW at a frequency of 920 kHz. Estimate the intensity of the radio wave at a point 10 km from the broadcast antenna.

65. ‖ LASIK eye surgery uses pulses of laser light to shave off
   BIO tissue from the cornea, reshaping it. A typical LASIK laser emits a 1.0-mm-diameter laser beam with a wavelength of 193 nm. Each laser pulse lasts 15 ns and contains 1.0 mJ of light energy
   a. What is the power of one laser pulse?
   b. During the very brief time of the pulse, what is the intensity of the light wave?

66. ‖ The sound intensity 50 m from a wailing tornado siren is 0.10 W/m².
   a. What is the intensity at 1000 m?
   b. The weakest intensity likely to be heard over background noise is $\approx 1\ \mu$W/m². Estimate the maximum distance at which the siren can be heard.

67. ‖ The sound intensity level 5.0 m from a large power saw is 100 dB. At what distance will the sound be a more tolerable 80 dB?

68. ‖ Two loudspeakers on elevated platforms are at opposite ends of a field. Each broadcasts equally in all directions. The sound intensity level at a point halfway between the loudspeakers is 75.0 dB. What is the sound intensity level at a point one-quarter of the way from one speaker to the other along the line joining them?

69. ‖ Your ears are sensitive to differences in pitch, but they are not
   BIO very sensitive to differences in intensity. You are not capable of detecting a difference in sound intensity level of less than 1 dB. By what factor does the sound intensity increase if the sound intensity level increases from 60 dB to 61 dB?

70. ‖‖ The intensity of a sound source is described by an inverse-square law only if the source is very small (a point source) and only if the waves can travel unimpeded in all directions. For an extended source or in a situation where obstacles absorb or reflect the waves, the intensity at distance r can often be expressed as $I = cP_{\text{source}}/r^x$, where c is a constant and the exponent x—which would be 2 for an ideal spherical wave—depends on the situation. In one such situation, you use a sound meter to measure the sound intensity level at different distances from a source, acquiring the following data:

Distance (m)	Intensity level (dB)
1	100
3	93
10	85
30	78
100	70

Use the best-fit line of an appropriate graph to determine the exponent x that characterizes this sound source.

71. ‖‖ A mad doctor believes that baldness can be cured by warming the scalp with sound waves. His patients sit underneath the Bald-o-Matic loudspeakers, where their heads are bathed with 93 dB of soothing 800 Hz sound waves. Suppose we model a bald head as a 16-cm-diameter hemisphere. If 0.10 J of sound energy is considered an appropriate "dose," how many minutes should each therapy session last?

72. ‖ A physics professor demonstrates the Doppler effect by tying a 600 Hz sound generator to a 1.0-m-long rope and whirling it around her head in a horizontal circle at 100 rpm. What are the highest and lowest frequencies heard by a student in the classroom?

73. ‖ Show that the Doppler frequency f_- of a receding source is $f_- = f_0/(1 + v_s/v)$.

74. ‖ A starship approaches its home planet at a speed of $0.1c$. When it is 54×10^6 km away, it uses its green laser beam ($\lambda = 540$ nm) to signal its approach.
   a. How long does the signal take to travel to the home planet?
   b. At what wavelength is the signal detected on the home planet?

75. ‖ Wavelengths of light from a distant galaxy are found to be 0.5% longer than the corresponding wavelengths measured in a terrestrial laboratory. Is the galaxy approaching or receding from the earth? At what speed?

76. ‖ You have just been pulled over for running a red light, and the police officer has informed you that the fine will be $250. In desperation, you suddenly recall an idea that your physics professor recently discussed in class. In your calmest voice, you tell the officer that the laws of physics prevented you from knowing that the light was red. In fact, as you drove toward it, the light was Doppler shifted to where it appeared green to you. "OK," says the officer, "Then I'll ticket you for speeding. The fine is $1 for every 1 km/h over the posted speed limit of 50 km/h." How big is your fine? Use 650 nm as the wavelength of red light and 540 nm as the wavelength of green light.

Challenge Problems

77. One way to monitor global warming is to measure the average temperature of the ocean. Researchers are doing this by measuring the time it takes sound pulses to travel underwater over large distances. At a depth of 1000 m, where ocean temperatures hold steady near 4°C, the average sound speed is 1480 m/s. It's known from laboratory measurements that the sound speed increases 4.0 m/s for every 1.0°C increase in temperature. In one experiment, where sounds generated near California are detected in the South Pacific, the sound waves travel 8000 km. If the smallest time change that can be reliably detected is 1.0 s, what is the smallest change in average temperature that can be measured?

78. The G string on a guitar is a 0.46-mm-diameter steel string with a linear density of 1.3 g/m. When the string is properly tuned to 196 Hz, the wave speed on the string is 250 m/s. Tuning is done by turning the tuning screw, which slowly tightens—and stretches—the string. By how many mm does a 75-cm-long G string stretch when it's first tuned?

79. A rope of mass m and length L hangs from a ceiling.
    a. Show that the wave speed on the rope a distance y above the lower end is $v = \sqrt{gy}$.
    b. Show that the time for a pulse to travel the length of the string is $\Delta t = 2\sqrt{L/g}$.

80. Some modern optical devices are made with glass whose index of refraction changes with distance from the front surface. FIGURE CP20.80 shows the index of refraction as a function of the distance into a slab of glass of thickness L. The index of refraction increases linearly from n_1 at the front surface to n_2 at the rear surface.

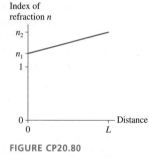

Index of refraction n

FIGURE CP20.80

a. Find an expression for the time light takes to travel through this piece of glass.
b. Evaluate your expression for a 1.0-cm-thick piece of glass for which $n_1 = 1.50$ and $n_2 = 1.60$.

81. A water wave is a *shallow-water wave* if the water depth d is less than $\approx \lambda/10$. It is shown in hydrodynamics that the speed of a shallow-water wave is $v = \sqrt{gd}$, so waves slow down as they move into shallower water. Ocean waves, with wavelengths of typically 100 m, are shallow-water waves when the water depth is less than ≈ 10 m. Consider a beach where the depth increases linearly with distance from the shore until reaching a depth of 5.0 m at a distance of 100 m. How long does it take a wave to move the last 100 m to the shore? Assume that the waves are so small that they don't break before reaching the shore.

82. An important characteristic of the heart, one used to diagnose heart disease, is the *pressure difference* between the blood pressure inside the heart and the blood pressure in the aorta, the large artery leading away from the heart. The blood inside the heart is essentially at rest, but it speeds up significantly as it enters the aorta—and its speed can be measured by using the Doppler shift of reflected ultrasound.
    a. The Doppler effect enters twice in calculating the frequency of the reflection from a moving object. Suppose the object's speed v_o is very small compared to the wave speed v. Show that a good approximation for the *Doppler shift*—the difference between the reflected frequency and the incident frequency—is

$$\Delta f = 2f_0 \frac{v_o}{v}$$

    b. A doctor using 2.5 MHz ultrasound measures a 6000 Hz Doppler shift as the ultrasound reflects from blood ejected from the heart into the aorta. What is the blood pressure difference, in mm of Hg, between the inside of the heart and the aorta? Assume the patient is lying down so that there is no height difference between the heart and the aorta. The density of blood is 1060 kg/m^3.

STOP TO THINK ANSWERS

Stop to Think 20.1: d and e. The wave speed depends on properties of the medium, not on how you generate the wave. For a string, $v = \sqrt{T_s/\mu}$. Increasing the tension or decreasing the linear density (lighter string) will increase the wave speed.

Stop to Think 20.2: b. The wave is traveling to the right at 2.0 m/s, so each point on the wave passes $x = 0$ m, the point of interest, 2.0 s before reaching $x = 4.0$ m. The graph has the same shape, but everything happens 2.0 s earlier.

Stop to Think 20.3: d. The wavelength—the distance between two crests—is seen to be 10 m. The frequency is $f = v/\lambda = (50 \text{ m/s})/(10 \text{ m}) = 5$ Hz.

Stop to Think 20.4: e. A crest and an adjacent trough are separated by $\lambda/2$. This is a phase difference of π rad.

Stop to Think 20.5: $n_c > n_a > n_b$. $\lambda = \lambda_{vac}/n$, so a shorter wavelength corresponds to a larger index of refraction.

Stop to Think 20.6: c. Any factor-of-2 change in intensity changes the sound intensity level by 3 dB. One trumpet is $\frac{1}{4}$ the original number, so the intensity has decreased by two factors of 2.

Stop to Think 20.7: c. Zack hears a higher frequency as he and the source approach. Amy is moving with the source, so $f_{Amy} = f_0$.

21 Superposition

This swirl of colors is due to a very thin layer of oil. Oil is clear. The colors arise from the interference of reflected light waves.

▶ **Looking Ahead** The goal of Chapter 21 is to understand and use the idea of superposition.

Standing Waves

Standing waves are created from the superposition of traveling waves bouncing back and forth between the edges of the medium.

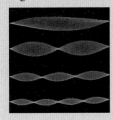

Standing waves occur in well-defined patterns called **modes,** each with its own distinct frequency. Some points on the wave, called **nodes,** do not oscillate at all.

You'll learn how to calculate the frequencies and wavelengths of standing waves on strings and in air.

Applications

You'll learn how standing waves determine the notes of a guitar and other musical instruments ...

...and how interference is used to design antireflection coatings for lenses.

◀ **Looking Back**
Section 20.5 Sound waves

Superposition

Waves can pass through each other—a characteristic that distinguishes waves from particles. As they do, their displacements add together. This is called the **principle of superposition**.

These water waves are exhibiting superposition as the ripples pass through each other.

You'll learn to analyze the response of the medium when two waves pass through each other.

◀ **Looking Back**
Sections 20.2–20.4 Properties of traveling waves

Interference

When two sources emit waves of the same wavelength and frequency, the overlapped waves create an **interference pattern.**

You'll learn to interpret interference patterns such as this one. The two black dots are the sources of the waves.

Constructive interference occurs where the waves add to make a larger wave. *Destructive interference* is where the waves cancel to make a smaller wave.

Beats

The superposition of two waves of slightly different frequency produces a soft-loud-soft-loud- ... modulation of the intensity called **beats.**

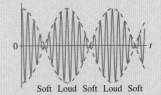

Soft Loud Soft Loud Soft

Beats are easily demonstrated with sound waves, but the concept is used in applications from ultrasonics to telecommunications.

21.1 The Principle of Superposition

FIGURE 21.1a shows two baseball players, Alan and Bill, at batting practice. Unfortunately, someone has turned the pitching machines so that pitching machine A throws baseballs toward Bill while machine B throws toward Alan. If two baseballs are launched at the same time, and with the same speed, they collide at the crossing point. Two particles cannot occupy the same point of space at the same time.

FIGURE 21.1 Unlike particles, two waves can pass directly through each other.

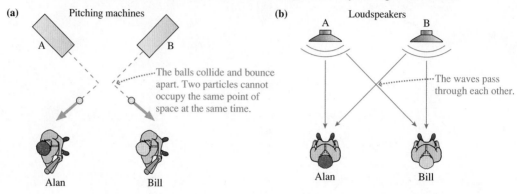

But waves, unlike particles, can pass directly through each other. In FIGURE 21.1b Alan and Bill are listening to the stereo system in the locker room after practice. Because both hear the music quite well, the sound wave that travels from loudspeaker A toward Bill must pass through the wave traveling from loudspeaker B toward Alan. What happens to the medium at a point where two waves are present simultaneously?

If wave 1 displaces a particle in the medium by D_1 and wave 2 *simultaneously* displaces it by D_2, the net displacement of the particle is simply $D_1 + D_2$. This is a very important idea because it tells us how to combine waves. It is known as the *principle of superposition.*

> **Principle of superposition** When two or more waves are *simultaneously* present at a single point in space, the displacement of the medium at that point is the sum of the displacements due to each individual wave.

When one object is placed on top of another, the two are said to be *superimposed.* But through some quirk in the English language, the result of superimposing objects is called a *superposition,* without the syllable "im." When one wave is "placed" on top of another wave, we have a superposition of waves.

Mathematically, the net displacement of a particle in the medium is

$$D_{net} = D_1 + D_2 + \cdots = \sum_i D_i \qquad (21.1)$$

where D_i is the displacement that would be caused by wave i alone. We will make the simplifying assumption that the displacements of the individual waves are along the same line so that we can add displacements as scalars rather than vectors.

To use the principle of superposition you must know the displacement caused by each wave if traveling alone. Then you go through the medium *point by point* and add the displacements due to each wave *at that point* to find the net displacement at that point. The outcome will be different at each and every point in the medium because the displacements are different at each point.

To illustrate, FIGURE 21.2 shows snapshot graphs taken 1 s apart of two waves traveling at the same speed (1 m/s) in opposite directions along a string. The principle of superposition comes into play wherever the waves overlap. The solid line is the sum *at each point* of the two displacements at that point. This is the displacement that you would actually observe as the two waves pass through each other.

FIGURE 21.2 The superposition of two waves on a string as they pass through each other.

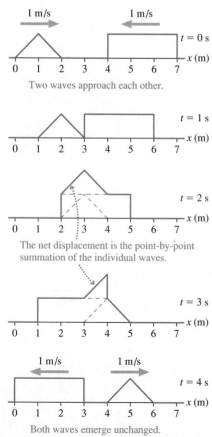

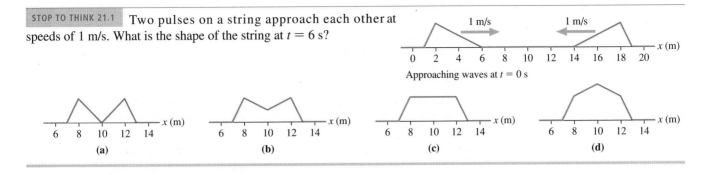

STOP TO THINK 21.1   Two pulses on a string approach each other at speeds of 1 m/s. What is the shape of the string at $t = 6$ s?

Approaching waves at $t = 0$ s

(a)   (b)   (c)   (d)

21.2 **Standing Waves**

FIGURE 21.3 is a time-lapse photograph of a *standing wave* on a vibrating string. It's not obvious from the photograph, but this is actually a superposition of two waves. To understand this, consider two sinusoidal waves **with the same frequency, wavelength, and amplitude** traveling in opposite directions. For example, **FIGURE 21.4a** shows two waves on a string, and **FIGURE 21.4b** shows nine snapshot graphs, at intervals of $\frac{1}{8}T$. The dots identify two of the crests to help you visualize the wave movement.

At *each point,* the net displacement—the superposition—is found by adding the red displacement and the green displacement. **FIGURE 21.4c** shows the result. It is the wave you would actually observe. The blue dot shows that the blue wave is moving neither right nor left. The wave of Figure 21.4c is called a **standing wave** because the crests and troughs "stand in place" as the wave oscillates.

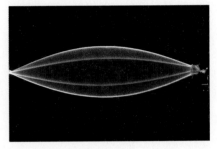

FIGURE 21.3 A vibrating string is an example of a standing wave.

FIGURE 21.4 The superposition of two sinusoidal waves traveling in opposite directions.

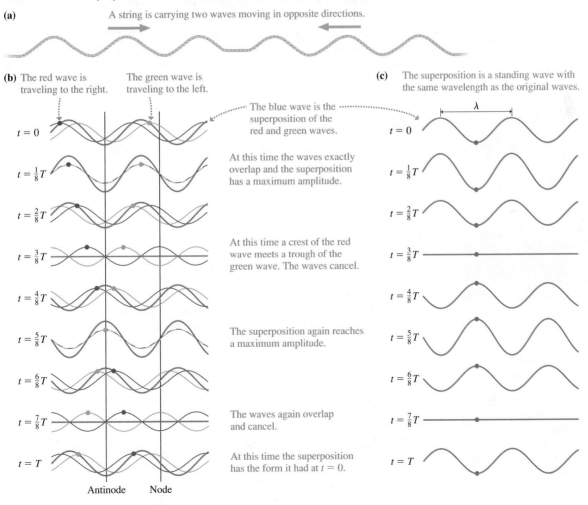

(a)   A string is carrying two waves moving in opposite directions.

(b) The red wave is traveling to the right.

The green wave is traveling to the left.

The blue wave is the superposition of the red and green waves.

$t = 0$

$t = \frac{1}{8}T$

At this time the waves exactly overlap and the superposition has a maximum amplitude.

$t = \frac{2}{8}T$

$t = \frac{3}{8}T$

At this time a crest of the red wave meets a trough of the green wave. The waves cancel.

$t = \frac{4}{8}T$

$t = \frac{5}{8}T$

The superposition again reaches a maximum amplitude.

$t = \frac{6}{8}T$

$t = \frac{7}{8}T$

The waves again overlap and cancel.

$t = T$

At this time the superposition has the form it had at $t = 0$.

Antinode   Node

(c)   The superposition is a standing wave with the same wavelength as the original waves.

$t = 0$ λ

$t = \frac{1}{8}T$

$t = \frac{2}{8}T$

$t = \frac{3}{8}T$

$t = \frac{4}{8}T$

$t = \frac{5}{8}T$

$t = \frac{6}{8}T$

$t = \frac{7}{8}T$

$t = T$

FIGURE 21.5 Standing waves are often represented as they would be seen in a time-lapse photograph.

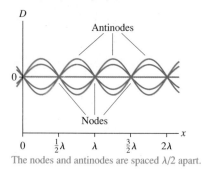

The nodes and antinodes are spaced λ/2 apart.

FIGURE 21.6 The intensity of a standing wave is maximum at the antinodes, zero at the nodes.

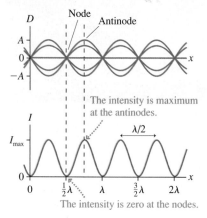

The intensity is maximum at the antinodes.

The intensity is zero at the nodes.

This photograph shows the Tacoma Narrows suspension bridge on the day in 1940 when it experienced a catastrophic standing-wave oscillation that led to its collapse. Aerodynamic forces caused the amplitude of a particular resonant mode of the bridge to increase dramatically until the bridge failed. In this photo, the red line shows the original line of the deck of the bridge. You can clearly see the large amplitude of the oscillation and the node at the center of the span.

Nodes and Antinodes

FIGURE 21.5 has collapsed the nine graphs of Figure 21.4b into a single graphical representation of a standing wave. Compare this to the Figure 21.3 photograph of a vibrating string. A striking feature of a standing-wave pattern is the existence of **nodes,** points that *never move!* **The nodes are spaced λ/2 apart.** Halfway between the nodes are the points where the particles in the medium oscillate with maximum displacement. These points of maximum amplitude are called **antinodes,** and you can see that they are also spaced λ/2 apart.

It seems surprising and counterintuitive that some particles in the medium have no motion at all. To understand how this happens, look carefully at the two traveling waves in Figure 21.4a. You will see that the nodes occur at points where at *every instant* of time the displacements of the two traveling waves have equal magnitudes but *opposite signs.* Thus the superposition of the displacements at these points is always zero. The antinodes correspond to points where the two displacements have equal magnitudes and the *same sign* at all times.

Two waves 1 and 2 are said to be *in phase* at a point where D_1 is *always* equal to D_2. The superposition at that point yields a wave whose amplitude is twice that of the individual waves. This is called a point of *constructive interference.* The antinodes of a standing wave are points of constructive interference between the two traveling waves.

In contrast, two waves are said to be *out of phase* at points where D_1 is *always* equal to $-D_2$. Their superposition gives a wave with zero amplitude—no wave at all! This is a point of *destructive interference.* The nodes of a standing wave are points of destructive interference. We will defer the main discussion of constructive and destructive interference until later in this chapter, but you'll then recognize that you're seeing constructive and destructive interference at the antinodes and nodes of a standing wave.

In Chapter 20 you learned that the *intensity* of a wave is proportional to the square of the amplitude: $I \propto A^2$. You can see in **FIGURE 21.6** that maximum intensity occurs at the antinodes and that the intensity is zero at the nodes. If this is a sound wave, the loudness is maximum at the antinodes and zero at the nodes. A standing light wave is bright at the antinodes, dark at the nodes. The key idea is that **the intensity is maximum at points of constructive interference and zero at points of destructive interference.**

The Mathematics of Standing Waves

A sinusoidal wave traveling to the right along the *x*-axis with angular frequency $\omega = 2\pi f$, wave number $k = 2\pi/\lambda$, and amplitude a is

$$D_R = a\sin(kx - \omega t) \tag{21.2}$$

An equivalent wave traveling to the left is

$$D_L = a\sin(kx + \omega t) \tag{21.3}$$

We previously used the symbol A for the wave amplitude, but here we will use a lowercase a to represent the amplitude of each individual wave and reserve A for the amplitude of the net wave.

According to the principle of superposition, the net displacement of the medium when both waves are present is the sum of D_R and D_L:

$$D(x, t) = D_R + D_L = a\sin(kx - \omega t) + a\sin(kx + \omega t) \tag{21.4}$$

We can simplify Equation 21.4 by using the trigonometric identity

$$\sin(\alpha \pm \beta) = \sin\alpha\cos\beta \pm \cos\alpha\sin\beta$$

Doing so gives

$$D(x, t) = a(\sin kx\cos\omega t - \cos kx\sin\omega t) + a(\sin kx\cos\omega t + \cos kx\sin\omega t)$$

$$= (2a\sin kx)\cos\omega t \tag{21.5}$$

It is useful to write Equation 21.5 as

$$D(x, t) = A(x) \cos \omega t \qquad (21.6)$$

where the **amplitude function** $A(x)$ is defined as

$$A(x) = 2a \sin kx \qquad (21.7)$$

The amplitude reaches a maximum value $A_{max} = 2a$ at points where $\sin kx = 1$.

The displacement $D(x, t)$ given by Equation 21.6 is neither a function of $x - vt$ nor a function of $x + vt$; hence it is *not* a traveling wave. Instead, the $\cos \omega t$ term in Equation 21.6 describes a medium in which each point oscillates in simple harmonic motion with frequency $f = \omega/2\pi$. The function $A(x) = 2a \sin kx$ gives the amplitude of the oscillation for a particle at position x.

FIGURE 21.7 graphs Equation 21.6 at several different instants of time. Notice that the graphs are identical to those of Figure 21.5, showing us that Equation 21.6 is the mathematical description of a standing wave.

The nodes of the standing wave are the points at which the amplitude is zero. They are located at positions x for which

$$A(x) = 2a \sin kx = 0 \qquad (21.8)$$

The sine function is zero if the angle is an integer multiple of π rad, so Equation 21.8 is satisfied if

$$kx_m = \frac{2\pi x_m}{\lambda} = m\pi \qquad m = 0, 1, 2, 3, \dots \qquad (21.9)$$

Thus the position x_m of the mth node is

$$x_m = m\frac{\lambda}{2} \qquad m = 0, 1, 2, 3, \dots \qquad (21.10)$$

You can see that the spacing between two adjacent nodes is $\lambda/2$, in agreement with Figure 21.6. The nodes are *not* spaced by λ, as you might have expected.

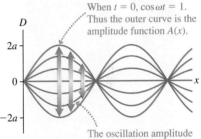

FIGURE 21.7 The net displacement resulting from two counter-propagating sinusoidal waves.

When $t = 0$, $\cos \omega t = 1$. Thus the outer curve is the amplitude function $A(x)$.

The oscillation amplitude changes with position.

EXAMPLE 21.1   **Node spacing on a string**

A very long string has a linear density of 5.0 g/m and is stretched with a tension of 8.0 N. 100 Hz waves with amplitudes of 2.0 mm are generated at the ends of the string.

a. What is the node spacing along the resulting standing wave?
b. What is the maximum displacement of the string?

MODEL   Two counter-propagating waves of equal frequency create a standing wave.

VISUALIZE   The standing wave will look like Figure 21.5.

SOLVE a.   The speed of the waves on the string is

$$v = \sqrt{\frac{T_s}{\mu}} = \sqrt{\frac{8.0\ \text{N}}{0.0050\ \text{kg/m}}} = 40\ \text{m/s}$$

and the wavelength is

$$\lambda = \frac{v}{f} = \frac{40\ \text{m/s}}{100\ \text{Hz}} = 0.40\ \text{m} = 40\ \text{cm}$$

Thus the spacing between adjacent nodes is $\lambda/2 = 20$ cm.
b. The maximum displacement is $A_{max} = 2a = 4.0$ mm.

21.3 Standing Waves on a String

Wiggling both ends of a very long string is not a practical way to generate standing waves. Instead, as in the photograph in Figure 21.3, standing waves are usually seen on a string that is fixed at both ends. To understand why this condition causes standing waves, we need to examine what happens when a traveling wave encounters a discontinuity.

FIGURE 21.8a on the next page shows a *discontinuity* between a string with a larger linear density and one with a smaller linear density. The tension is the same in both strings, so the wave speed is slower on the left, faster on the right. Whenever a wave encounters a discontinuity, some of the wave's energy is *transmitted* forward and some is *reflected*.

FIGURE 21.8 A wave reflects when it encounters a discontinuity or a boundary.

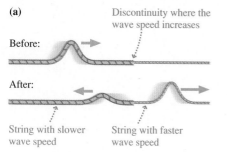

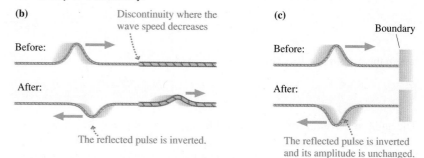

(a)

Before:

Discontinuity where the wave speed increases

After:

String with slower wave speed · String with faster wave speed

(b)

Before:

Discontinuity where the wave speed decreases

After:

The reflected pulse is inverted.

(c)

Boundary

Before:

After:

The reflected pulse is inverted and its amplitude is unchanged.

FIGURE 21.9 A strobe photo of a pulse traveling along a rope-like spring.

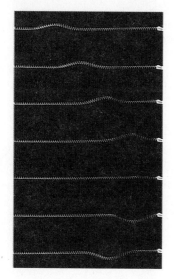

FIGURE 21.10 Reflections at the two boundaries cause a standing wave on the string.

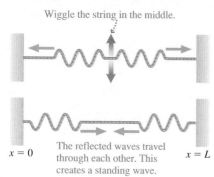

Wiggle the string in the middle.

$x = 0$ The reflected waves travel through each other. This creates a standing wave. $x = L$

Light waves exhibit an analogous behavior when they encounter a piece of glass. Most of the light wave's energy is transmitted through the glass, which is why glass is transparent, but a small amount of energy is reflected. That is how you see your reflection dimly in a storefront window.

In **FIGURE 21.8b**, an incident wave encounters a discontinuity at which the wave speed decreases. In this case, the reflected pulse is *inverted*. A positive displacement of the incident wave becomes a negative displacement of the reflected wave. Because $\sin(\phi + \pi) = -\sin\phi$, we say that the reflected wave has a *phase change of π upon reflection*. This aspect of reflection will be important later in the chapter when we look at the interference of light waves.

The wave in **FIGURE 21.8c** reflects from a *boundary*. You can think of this as Figure 21.8b in the limit that the string on the right becomes infinitely massive. Thus the reflection in Figure 21.8c looks like that of Figure 21.8b with one exception: Because there is no transmitted wave, *all* the wave's energy is reflected. Hence **the amplitude of a wave reflected from a boundary is unchanged.** **FIGURE 21.9** is a sequence of strobe photos in which you see a pulse on a rope-like spring reflecting from a boundary at the right of the photo. The reflected pulse is inverted but otherwise unchanged.

Creating Standing Waves

FIGURE 21.10 shows a string of length L tied at $x = 0$ and $x = L$. If you wiggle the string in the middle, sinusoidal waves travel outward in both directions and soon reach the boundaries. Because the speed of a reflected wave does not change, **the wavelength and frequency of a reflected sinusoidal wave are unchanged.** Consequently, reflections at the ends of the string cause two waves of *equal amplitude and wavelength* to travel in opposite directions along the string. As we've just seen, these are the conditions that cause a standing wave!

To connect the mathematical analysis of standing waves in Section 21.2 with the physical reality of a string tied down at the ends, we need to impose *boundary conditions*. A **boundary condition** is a mathematical statement of any constraint that *must* be obeyed at the boundary or edge of a medium. Because the string is tied down at the ends, the displacements at $x = 0$ and $x = L$ must be zero at all times. Thus the standing-wave boundary conditions are $D(x = 0, t) = 0$ and $D(x = L, t) = 0$. Stated another way, we require nodes at both ends of the string.

We found that the displacement of a standing wave is $D(x, t) = (2a \sin kx) \cos \omega t$. This equation already satisfies the boundary condition $D(x = 0, t) = 0$. That is, the origin has already been located at a node. The second boundary condition, at $x = L$, requires $D(x = L, t) = 0$. This condition will be met at all times if

$$2a \sin kL = 0 \qquad \text{(boundary condition at } x = L\text{)} \qquad (21.11)$$

Equation 21.11 will be true if $\sin kL = 0$, which in turn requires

$$kL = \frac{2\pi L}{\lambda} = m\pi \qquad m = 1, 2, 3, 4, \ldots \qquad (21.12)$$

kL must be a multiple of $m\pi$, but $m = 0$ is excluded because L can't be zero.

For a string of fixed length L, the only quantity in Equation 21.12 that can vary is λ. That is, the boundary condition is satisfied only if the wavelength has one of the values

$$\lambda_m = \frac{2L}{m} \qquad m = 1, 2, 3, 4, \ldots \qquad (21.13)$$

A standing wave can exist on the string *only* if its wavelength is one of the values given by Equation 21.13. The mth possible wavelength $\lambda_m = 2L/m$ is just the right size so that its mth node is located at the end of the string (at $x = L$).

NOTE ▶ Other wavelengths, which would be perfectly acceptable wavelengths for a traveling wave, cannot exist as a *standing* wave of length L because they cannot meet the boundary conditions requiring a node at each end of the string. ◀

If standing waves are possible only for certain wavelengths, then only a few specific oscillation frequencies are allowed. Because $\lambda f = v$ for a sinusoidal wave, the oscillation frequency corresponding to wavelength λ_m is

$$f_m = \frac{v}{\lambda_m} = \frac{v}{2L/m} = m\frac{v}{2L} \qquad m = 1, 2, 3, 4, \ldots \qquad (21.14)$$

The lowest allowed frequency

$$f_1 = \frac{v}{2L} \qquad \text{(fundamental frequency)} \qquad (21.15)$$

which corresponds to wavelength $\lambda_1 = 2L$, is called the **fundamental frequency** of the string. The allowed frequencies can be written in terms of the fundamental frequency as

$$f_m = mf_1 \qquad m = 1, 2, 3, 4, \ldots \qquad (21.16)$$

The allowed standing-wave frequencies are all integer multiples of the fundamental frequency. The higher-frequency standing waves are called **harmonics,** with the $m = 2$ wave at frequency f_2 called the *second harmonic,* the $m = 3$ wave called the *third harmonic,* and so on.

FIGURE 21.11 graphs the first four possible standing waves on a string of fixed length L. These possible standing waves are called the **modes** of the string, or sometimes the *normal modes.* Each mode, numbered by the integer m, has a unique wavelength and frequency. Keep in mind that these drawings simply show the *envelope,* or outer edge, of the oscillations. The string is continuously oscillating at all positions between these edges, as we showed in more detail in Figure 21.5.

There are three things to note about the modes of a string.

1. m is the number of *antinodes* on the standing wave, not the number of nodes. You can tell a string's mode of oscillation by counting the number of antinodes.
2. The *fundamental mode,* with $m = 1$, has $\lambda_1 = 2L$, not $\lambda_1 = L$. Only half of a wavelength is contained between the boundaries, a direct consequence of the fact that the spacing between nodes is $\lambda/2$.
3. The frequencies of the normal modes form a series: $f_1, 2f_1, 3f_1, 4f_1, \ldots$. The fundamental frequency f_1 can be found as the *difference* between the frequencies of any two adjacent modes. That is, $f_1 = \Delta f = f_{m+1} - f_m$.

FIGURE 21.12 is a time-exposure photograph of the $m = 3$ standing wave on a string. The nodes and antinodes are quite distinct. The string vibrates three times faster for the $m = 3$ mode than for the fundamental $m = 1$ mode.

FIGURE 21.11 The first four modes for standing waves on a string of length L.

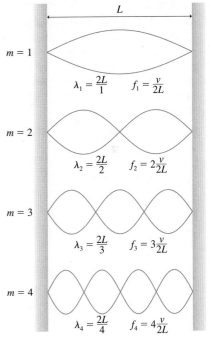

FIGURE 21.12 Time-exposure photograph of the $m = 3$ standing-wave mode on a stretched string.

EXAMPLE 21.2 **Measuring *g***

Standing-wave frequencies can be measured very accurately. Consequently, standing waves are often used in experiments to make accurate measurements of other quantities. One such experiment, shown in FIGURE 21.13, uses standing waves to measure the free-fall acceleration *g*. A heavy mass is suspended from a 1.65-m-long, 5.85 g steel wire; then an oscillating magnetic field (because steel is magnetic) is used to excite the *m* = 3 standing wave on the wire. Measuring the frequency for different masses yields the following data:

FIGURE 21.13 An experiment to measure *g* with standing waves.

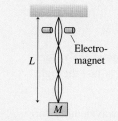

Mass (kg)	f_3 (Hz)
2.00	68
4.00	97
6.00	117
8.00	135
10.00	152

Analyze these data to determine the local value of *g*.

MODEL The hanging mass creates tension in the wire. This establishes the wave speed along the wire and thus the frequencies of standing waves. Masses of a few kg might stretch the wire a mm or so, but that doesn't change the length *L* until the third decimal place. The mass of the wire itself is insignificant in comparison to that of the hanging mass. We'll be justified in determining *g* to three significant figures.

SOLVE The frequency of the third harmonic is

$$f_3 = \frac{3}{2}\frac{v}{L}$$

The wave speed on the wire is

$$v = \sqrt{\frac{T_s}{\mu}} = \sqrt{\frac{Mg}{m/L}} = \sqrt{\frac{MgL}{m}}$$

where *Mg* is the weight of the hanging mass, and thus the tension in the wire, while *m* is the mass of the wire. Combining these two equations, we have

$$f_3 = \frac{3}{2}\sqrt{\frac{Mg}{mL}} = \frac{3}{2}\sqrt{\frac{g}{mL}}\sqrt{M}$$

Squaring both sides gives

$$f_3^2 = \frac{9g}{4mL}M$$

A graph of the square of the standing-wave frequency versus mass *M* should be a straight line passing through the origin with slope $9g/4mL$. We can use the experimental slope to determine *g*. FIGURE 21.14 is a graph of f_3^2 versus *M*. The slope of the best-fit line is 2289 kg^{-1}s^{-2}, from which we find

$$g = \text{slope} \times \frac{4mL}{9}$$

$$= 2289 \text{ kg}^{-1}\text{s}^{-2} \times \frac{4(0.00585 \text{ kg})(1.65 \text{ m})}{9} = 9.82 \text{ m/s}^2$$

FIGURE 21.14 Graph of the data.

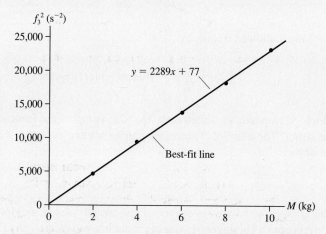

ASSESS The fact that the graph is linear and passes through the origin confirms our model. This is an important reason for having multiple data points rather than using only one mass.

STOP TO THINK 21.2 A standing wave on a string vibrates as shown at the right. Suppose the string tension is quadrupled while the frequency and the length of the string are held constant. Which standing-wave pattern is produced?

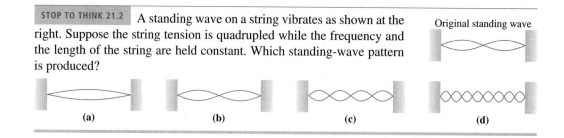

Standing Electromagnetic Waves

Because electromagnetic waves are transverse waves, a standing electromagnetic wave is very much like a standing wave on a string. Standing electromagnetic waves can be established between two parallel mirrors that reflect light back and forth. The mirrors are boundaries, analogous to the boundaries at the ends of a string. In fact, this is exactly how a laser operates. The two facing mirrors in FIGURE 21.15 form what is called a *laser cavity*.

Because the mirrors act like the points to which a string is tied, the light wave must have a node at the surface of each mirror. One of the mirrors is only partially reflective, to allow some light to escape and form the laser beam, but this doesn't affect the boundary condition.

Because the boundary conditions are the same, Equations 21.13 and 21.14 for λ_m and f_m apply to a laser just as they do to a vibrating string. The primary difference is the size of the wavelength. A typical laser cavity has a length $L \approx 30$ cm, and visible light has a wavelength $\lambda \approx 600$ nm. The standing light wave in a laser cavity has a mode number m that is approximately

$$m = \frac{2L}{\lambda} \approx \frac{2 \times 0.30 \text{ m}}{6.00 \times 10^{-7} \text{ m}} = 1{,}000{,}000$$

In other words, the standing light wave inside a laser cavity has approximately one million antinodes! This is a consequence of the very short wavelength of light.

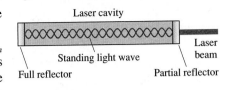

FIGURE 21.15 A laser contains a standing light wave between two parallel mirrors.

Laser cavity

Standing light wave

Laser beam

Full reflector   Partial reflector

EXAMPLE 21.3 | **The standing light wave inside a laser**

Helium-neon lasers emit the red laser light commonly used in classroom demonstrations and supermarket checkout scanners. A helium-neon laser operates at a wavelength of precisely 632.9924 nm when the spacing between the mirrors is 310.372 mm.

a. In which mode does this laser operate?
b. What is the next longest wavelength that could form a standing wave in this laser cavity?

MODEL The light wave forms a standing wave between the two mirrors.

VISUALIZE The standing wave looks like Figure 21.15.

SOLVE a. We can use $\lambda_m = 2L/m$ to find that m (the mode) is

$$m = \frac{2L}{\lambda_m} = \frac{2(0.310372 \text{ m})}{6.329924 \times 10^{-7} \text{ m}} = 980{,}650$$

There are 980,650 antinodes in the standing light wave.

b. The next longest wavelength that can fit in this laser cavity will have one fewer node. It will be the $m = 980{,}649$ mode and its wavelength will be

$$\lambda = \frac{2L}{m} = \frac{2(0.310372 \text{ m})}{980{,}649} = 632.9930 \text{ nm}$$

ASSESS The wavelength increases by a mere 0.0006 nm when the mode number is decreased by 1.

Microwaves, with a wavelength of a few centimeters, can also set up standing waves. This is not always good. If the microwaves in a microwave oven form a standing wave, there are nodes where the electromagnetic field intensity is always zero. These nodes cause cold spots where the food does not heat. Although designers of microwave ovens try to prevent standing waves, ovens usually do have cold spots spaced $\lambda/2$ apart at nodes in the microwave field. A turntable in a microwave oven keeps the food moving so that no part of your dinner remains at a node.

21.4 Standing Sound Waves and Musical Acoustics

A long, narrow column of air, such as the air in a tube or pipe, can support a *longitudinal* standing sound wave. Longitudinal waves are somewhat trickier than string waves because a graph—showing displacement *parallel* to the tube—is not a picture of the wave.

To illustrate the ideas, FIGURE 21.16 on the next page is a series of three graphs and pictures that show the $m = 2$ standing wave inside a column of air closed at both ends. We call this a *closed-closed tube*. The air at the closed ends cannot oscillate because the air molecules are pressed up against the wall, unable to move; hence **a closed end of a column of air must be a displacement node.** Thus the boundary conditions—nodes at the ends—are the same as for a standing wave on a string.

Although the graph looks familiar, it is now a graph of *longitudinal* displacement. At $t = 0$, positive displacements in the left half and negative displacements in the right half cause all the air molecules to converge at the center of the tube. The density and pressure rise at the center and fall at the ends—a *compression* and *rarefaction* in the terminology of Chapter 20. A half cycle later, the molecules have rushed to the ends

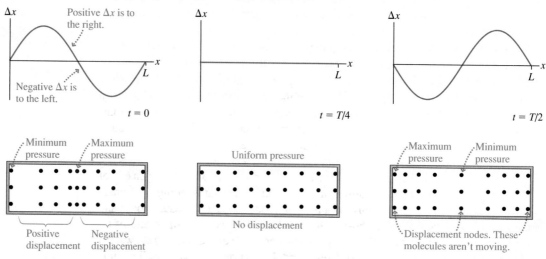

FIGURE 21.16 This time sequence of graphs and pictures illustrates the $m = 2$ standing sound wave in a closed-closed tube of air.

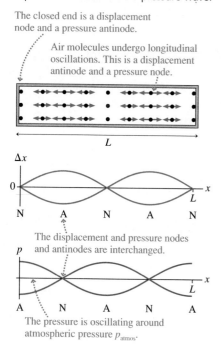

FIGURE 21.17 The $m = 2$ longitudinal standing wave can be represented as a displacement wave or as a pressure wave.

The closed end is a displacement node and a pressure antinode.

Air molecules undergo longitudinal oscillations. This is a displacement antinode and a pressure node.

The displacement and pressure nodes and antinodes are interchanged.

The pressure is oscillating around atmospheric pressure p_{atmos}.

of the tube. Now the pressure is maximum at the ends, minimum in the center. Try to visualize the air molecules sloshing back and forth this way.

FIGURE 21.17 combines these illustrations into a single picture showing where the molecules are oscillating (antinodes) and where they're not (nodes). A graph of the displacement Δx looks just like the $m = 2$ graph of a standing wave on a string. Because the boundary conditions are the same, the possible wavelengths and frequencies of standing waves in a closed-closed tube are the same as for a string of the same length.

It is often useful to think of sound as a *pressure wave* rather than a displacement wave, and the bottom graph in Figure 21.17 shows the $m = 2$ pressure standing wave in a closed-closed tube. Notice that the pressure is oscillating around p_{atmos}, its equilibrium value. **The nodes and antinodes of the pressure wave are interchanged with those of the displacement wave,** and a careful study of Figure 21.16 reveals why. The gas molecules are alternately pushed up against the ends of the tube, then pulled away, causing the pressure at the closed ends to oscillate with maximum amplitude—an antinode.

EXAMPLE 21.4 **Singing in the shower**

A shower stall is 2.45 m (8 ft) tall. For what frequencies less than 500 Hz are there standing sound waves in the shower stall?

MODEL The shower stall, to a first approximation, is a column of air 2.45 m long. It is closed at the ends by the ceiling and floor. Assume a 20°C speed of sound.

VISUALIZE A standing sound wave will have nodes at the ceiling and the floor. The $m = 2$ mode will look like Figure 21.17 rotated 90°.

SOLVE The fundamental frequency for a standing sound wave in this air column is

$$f_1 = \frac{v}{2L} = \frac{343 \text{ m/s}}{2(2.45 \text{ m})} = 70 \text{ Hz}$$

The possible standing-wave frequencies are integer multiples of the fundamental frequency. These are 70 Hz, 140 Hz, 210 Hz, 280 Hz, 350 Hz, 420 Hz, and 490 Hz.

ASSESS The many possible standing waves in a shower cause the sound to *resonate*, which helps explain why some people like to sing in the shower. Our approximation of the shower stall as a one-dimensional tube is actually a bit too simplistic. A three-dimensional analysis would find additional modes, making the "sound spectrum" even richer.

Air columns closed at both ends are of limited interest unless, as in Example 21.4, you are inside the column. Columns of air that *emit* sound are open at one or both ends. Many musical instruments fit this description. For example, a flute is a tube of air open at both ends. The flutist blows across one end to create a standing wave inside the tube, and a note of that frequency is emitted from both ends of the flute. (The blown end of a flute is open on the side, rather than across the tube. That is necessary for practical reasons of how flutes are played, but from a physics perspective this is the "end" of the tube because it opens the tube to the atmosphere.) A trumpet, however, is open at the bell end but is *closed* by the player's lips at the other end.

You saw earlier that a wave is partially transmitted and partially reflected at a discontinuity. When a sound wave traveling through a tube of air reaches an open end, some of the wave's energy is transmitted out of the tube to become the sound that you hear and some portion of the wave is reflected back into the tube. These reflections, analogous to the reflection of a string wave from a boundary, allow standing sound waves to exist in a tube of air that is open at one or both ends.

Not surprisingly, the *boundary condition* at the open end of a column of air is not the same as the boundary condition at a closed end. The air pressure at the open end of a tube is constrained to match the atmospheric pressure of the surrounding air. Consequently, the open end of a tube must be a pressure node. Because pressure nodes and antinodes are interchanged with those of the displacement wave, **an open end of an air column is required to be a displacement antinode.** (A careful analysis shows that the antinode is actually just outside the open end, but for our purposes we'll assume the antinode is exactly at the open end.)

FIGURE 21.18 shows displacement and pressure graphs of the first three standing-wave modes of a tube closed at both ends (a *closed-closed tube*), a tube open at both ends (an *open-open tube*), and a tube open at one end but closed at the other (an *open-closed tube*), all with the same length L. Notice the pressure and displacement boundary conditions. The standing wave in the open-open tube looks like the closed-closed tube except that the positions of the nodes and antinodes are interchanged. In both cases there are m half-wavelength segments between the ends; thus the wavelengths and frequencies of an open-open tube and a closed-closed tube are the same as those of a string tied at both ends:

FIGURE 21.18 The first three standing sound wave modes in columns of air with different boundary conditions.

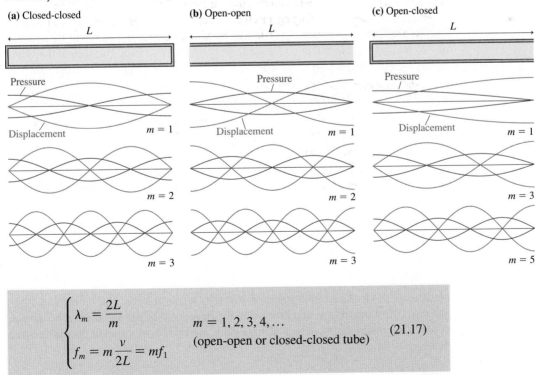

$$\begin{cases} \lambda_m = \dfrac{2L}{m} \\ f_m = m\dfrac{v}{2L} = mf_1 \end{cases} \quad \begin{array}{l} m = 1, 2, 3, 4, \ldots \\ \text{(open-open or closed-closed tube)} \end{array} \quad (21.17)$$

The open-closed tube is different. The fundamental mode has only one-quarter of a wavelength in a tube of length L; hence the $m = 1$ wavelength is $\lambda_1 = 4L$. This is twice the λ_1 wavelength of an open-open or a closed-closed tube. Consequently, **the fundamental frequency of an open-closed tube is half that of an open-open or a closed-closed tube of the same length.** It will be left as a homework problem for you to show that the possible wavelengths and frequencies of an open-closed tube of length L are

$$\begin{cases} \lambda_m = \dfrac{4L}{m} \\[2mm] f_m = m\dfrac{v}{4L} = mf_1 \end{cases} \quad \begin{array}{l} m = 1, 3, 5, 7, \dots \\ \text{(open-closed tube)} \end{array} \qquad (21.18)$$

Notice that m in Equation 21.18 takes on only *odd* values.

EXAMPLE 21.5 | Resonances of the ear canal

The eardrum, which transmits sounds vibrations to the sensory organs of the inner ear, lies at the end of the ear canal. For adults, the ear canal is about 2.5 cm in length. What frequency standing waves can occur in the ear canal that are within the range of human hearing? The speed of sound in the warm air of the ear canal is 350 m/s.

MODEL The ear canal is open to the air at one end, closed by the eardrum at the other. We can model it as an open-closed tube. The standing waves will be those of Figure 21.18c.

SOLVE The lowest standing-wave frequency is the fundamental frequency for a 2.5-cm-long open-closed tube:

$$f_1 = \frac{v}{4L} = \frac{350 \text{ m/s}}{4(0.025 \text{ m})} = 3500 \text{ Hz}$$

Standing waves also occur at the harmonics, but an open-closed tube has only odd harmonics. These are

$$f_3 = 3f_1 = 10{,}500 \text{ Hz}$$
$$f_5 = 5f_1 = 17{,}500 \text{ Hz}$$

Higher harmonics are beyond the range of human hearing, as discussed in Section 20.5.

ASSESS The ear canal is short, so we expected the standing-wave frequencies to be relatively high. The air in your ear canal responds readily to sounds at these frequencies—what we call a *resonance* of the ear canal—and transmits theses sounds to the eardrum. Consequently, your ear actually is slightly more sensitive to sounds with frequencies around 3500 Hz and 10,500 Hz than to sounds at nearby frequencies.

STOP TO THINK 21.3 An open-open tube of air supports standing waves at frequencies of 300 Hz and 400 Hz and at no frequencies between these two. The second harmonic of this tube has frequency

a. 100 Hz   b. 200 Hz   c. 400 Hz   d. 600 Hz   e. 800 Hz

Musical Instruments

An important application of standing waves is to musical instruments. Instruments such as the guitar, the piano, and the violin have strings fixed at the ends and tightened to create tension. A disturbance generated on the string by plucking, striking, or bowing it creates a standing wave on the string.

The fundamental frequency of a vibrating string is

$$f_1 = \frac{v}{2L} = \frac{1}{2L}\sqrt{\frac{T_s}{\mu}}$$

where T_s is the tension in the string and μ is its linear density. The fundamental frequency is the musical note you hear when the string is sounded. Increasing the tension in the string raises the fundamental frequency, which is how stringed instruments are tuned.

NOTE ▶ v is the wave speed *on the string,* not the speed of sound in air. ◀

For the guitar or the violin, the strings are all the same length and under approximately the same tension. Were that not the case, the neck of the instrument would tend to twist toward the side of higher tension. The strings have different frequencies because they differ in linear density: The lower-pitched strings are "fat" while the higher-pitched strings are "skinny." This difference changes the frequency by changing the wave speed. *Small* adjustments are then made in the tension to bring each string to the exact desired frequency. Once the instrument is tuned, you play it by using your fingertips to alter the effective length of the string. As you shorten the string's length, the frequency and pitch go up.

A piano covers a much wider range of frequencies than a guitar or violin. This range cannot be produced by changing only the linear densities of the strings. The high end would have strings too thin to use without breaking, and the low end would have solid rods rather than flexible wires! So a piano is tuned through a combination of changing the linear density *and* the length of the strings. The bass note strings are not only fatter, they are also longer.

With a wind instrument, blowing into the mouthpiece creates a standing sound wave inside a tube of air. The player changes the notes by using her fingers to cover holes or open valves, changing the length of the tube and thus its frequency. The fact that the holes are on the side makes very little difference; the first open hole becomes an antinode because the air is free to oscillate in and out of the opening.

A wind instrument's frequency depends on the speed of sound *inside* the instrument. But the speed of sound depends on the temperature of the air. When a wind player first blows into the instrument, the air inside starts to rise in temperature. This increases the sound speed, which in turn raises the instrument's frequency for each note until the air temperature reaches a steady state. Consequently, wind players must "warm up" before tuning their instrument.

Many wind instruments have a "buzzer" at one end of the tube, such as a vibrating reed on a saxophone or vibrating lips on a trombone. Buzzers generate a continuous range of frequencies rather than single notes, which is why they sound like a "squawk" if you play on just the mouthpiece without the rest of the instrument. When a buzzer is connected to the body of the instrument, most of those frequencies cause no response of the air molecules. But the frequency from the buzzer that matches the fundamental frequency of the instrument causes the buildup of a large-amplitude response at just that frequency—a standing-wave resonance. This is the energy input that generates and sustains the musical note.

The strings on a harp vibrate as standing waves. Their frequencies determine the notes that you hear.

EXAMPLE 21.6 **Flutes and clarinets**

A clarinet is 66.0 cm long. A flute is nearly the same length, with 63.5 cm between the hole the player blows across and the end of the flute. What are the frequencies of the lowest note and the next higher harmonic on a flute and on a clarinet? The speed of sound in warm air is 350 m/s.

MODEL The flute is an open-open tube, open at the end as well as at the hole the player blows across. A clarinet is an open-closed tube because the player's lips and the reed seal the tube at the upper end.

SOLVE The lowest frequency is the fundamental frequency. For the flute, an open-open tube, this is

$$f_1 = \frac{v}{2L} = \frac{350 \text{ m/s}}{2(0.635 \text{ m})} = 275 \text{ Hz}$$

The clarinet, an open-closed tube, has

$$f_1 = \frac{v}{4L} = \frac{350 \text{ m/s}}{4(0.660 \text{ m})} = 133 \text{ Hz}$$

The next higher harmonic on the flute's open-open tube is $m = 2$ with frequency $f_2 = 2f_1 = 550$ Hz. An open-closed tube has only odd harmonics, so the next higher harmonic of the clarinet is $f_3 = 3f_1 = 399$ Hz.

ASSESS The clarinet plays a much lower note than the flute—musically, about an octave lower—because it is an open-closed tube. It's worth noting that neither of our fundamental frequencies is exactly correct because our open-open and open-closed tube models are a bit too simplified to adequately describe a real instrument. However, both calculated frequencies are close because our models do capture the essence of the physics.

A vibrating string plays the musical note corresponding to the fundamental frequency f_1, so stringed instruments must use several strings to obtain a reasonable range of notes. In contrast, wind instruments can sound at the second or third harmonic of the tube of air (f_2 or f_3). These higher frequencies are sounded by *overblowing* (flutes, brass instruments) or with keys that open small holes to encourage the formation of an antinode at that point (clarinets, saxophones). The controlled use of these higher harmonics gives wind instruments a wide range of notes.

21.5 Interference in One Dimension

One of the most basic characteristics of waves is the ability of two waves to combine into a single wave whose displacement is given by the principle of superposition. The pattern resulting from the superposition of two waves is often called **interference.** A standing wave is the interference pattern produced when two waves of equal frequency travel in opposite directions. In this section we will look at the interference of two waves traveling in the *same* direction.

FIGURE 21.19 Two overlapped waves travel along the x-axis.

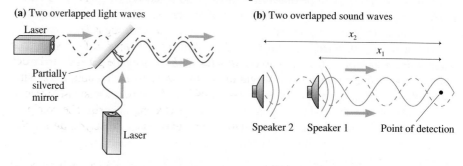

(a) Two overlapped light waves

Laser

Partially silvered mirror

Laser

(b) Two overlapped sound waves

x_2

x_1

Speaker 2   Speaker 1   Point of detection

FIGURE 21.19a shows two light waves impinging on a partially silvered mirror. Such a mirror partially transmits and partially reflects each wave, causing two *overlapped* light waves to travel along the x-axis to the right of the mirror. Or consider the two loudspeakers in **FIGURE 21.19b**. The sound wave from loudspeaker 2 passes just to the side of loudspeaker 1; hence two overlapped sound waves travel to the right along the x-axis. We want to find out what happens when two overlapped waves travel in the same direction along the same axis.

Figure 21.19b shows a point on the x-axis where the overlapped waves are detected, either by your ear or by a microphone. This point is distance x_1 from loudspeaker 1 and distance x_2 from loudspeaker 2. (We will use loudspeakers and sound waves for most of our examples, but our analysis is valid for any wave.) What is the amplitude of the combined waves at this point?

Throughout this section, **we will assume that the waves are sinusoidal, have the same frequency and amplitude, and travel to the right along the x-axis.** Thus we can write the displacements of the two waves as

$$D_1(x_1, t) = a\sin(kx_1 - \omega t + \phi_{10}) = a\sin\phi_1$$
$$D_2(x_2, t) = a\sin(kx_2 - \omega t + \phi_{20}) = a\sin\phi_2$$

(21.19)

where ϕ_1 and ϕ_2 are the *phases* of the waves. Both waves have the same wave number $k = 2\pi/\lambda$ and the same angular frequency $\omega = 2\pi f$.

The phase constants ϕ_{10} and ϕ_{20} are characteristics of *the sources,* not the medium. **FIGURE 21.20** shows snapshot graphs at $t = 0$ of waves emitted by three sources with phase constants $\phi_0 = 0$ rad, $\phi_0 = \pi/2$ rad, and $\phi_0 = \pi$ rad. You can see that **the phase constant tells us what the source is doing at $t = 0$.** For example, a loudspeaker at its center position and moving backward at $t = 0$ has $\phi_0 = 0$ rad. Looking back at Figure 21.19b, you can see that loudspeaker 1 has phase constant $\phi_{10} = 0$ rad and loudspeaker 2 has $\phi_{20} = \pi$ rad.

> **NOTE** ▶ We will often consider *identical sources,* by which we mean that $\phi_{20} = \phi_{10}$. That is, the sources oscillate in phase. ◀

Let's examine overlapped waves graphically before diving into the mathematics. **FIGURE 21.21** shows two important situations. In part a, the crests of the two waves are aligned as they travel along the x-axis. In part b, the crests of one wave align with the troughs of the other wave. The graphs and the wave fronts are slightly displaced from

FIGURE 21.20 Waves from three sources having phase constants $\phi_0 = 0$ rad, $\phi_0 = \pi/2$ rad, and $\phi_0 = \pi$ rad.

(a) Snapshot graph at $t = 0$ for $\phi_0 = 0$ rad

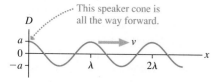

When this crest was emitted, a quarter cycle ago, the speaker cone was all the way forward.

D

a

0

$-a$

λ 2λ x

v

Now this speaker cone, at $x = 0$, is centered and moving backward.

(b) Snapshot graph at $t = 0$ for $\phi_0 = \pi/2$ rad

This speaker cone is all the way forward.

D

a

0

$-a$

λ 2λ x

v

(c) Snapshot graph at $t = 0$ for $\phi_0 = \pi$ rad

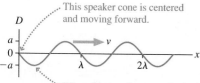

This speaker cone is centered and moving forward.

D

a

0

$-a$

λ 2λ x

v

When this trough was emitted, a quarter cycle ago, the speaker cone was all the way back.

each other so that you can see what each wave is doing, but the *physical situation* is one in which the waves are traveling *on top of* each other. Recall, from Chapter 20, that the wave fronts shown in the middle panel locate the crests of the waves.

The two waves of FIGURE 21.21a have the same displacement at every point: $D_1(x) = D_2(x)$. Two waves that are aligned crest to crest and trough to trough are said to be **in phase**. Waves that are in phase march along "in step" with each other.

When we combine two in-phase waves, using the principle of superposition, the net displacement at each point is twice the displacement of each individual wave. The superposition of two waves to create a traveling wave with an amplitude *larger* than either individual wave is called **constructive interference**. When the waves are exactly in phase, giving $A = 2a$, we have *maximum constructive interference*.

In FIGURE 21.21b, where the crests of one wave align with the troughs of the other, the waves march along "out of step" with $D_1(x) = -D_2(x)$ at every point. Two waves that are aligned crest to trough are said to be *180° out of phase* or, more generally, just **out of phase**. A superposition of two waves to create a wave with an amplitude smaller than either individual wave is called **destructive interference**. In this case, because $D_1 = -D_2$, the net displacement is *zero* at *every point* along the axis. The combination of two waves that cancel each other to give no wave is called *perfect destructive interference*.

NOTE ▶ Perfect destructive interference occurs only if the two waves have equal wavelengths and amplitudes, as we're assuming. Two waves of unequal amplitudes can interfere destructively, but the cancellation won't be perfect. ◀

The Phase Difference

To understand interference, we need to focus on the *phases* of the two waves, which are

$$\phi_1 = kx_1 - \omega t + \phi_{10}$$
$$\phi_2 = kx_2 - \omega t + \phi_{20} \qquad (21.20)$$

The difference between the two phases ϕ_1 and ϕ_2, called the **phase difference** $\Delta\phi$, is

$$\Delta\phi = \phi_2 - \phi_1 = (kx_2 - \omega t + \phi_{20}) - (kx_1 - \omega t + \phi_{10})$$
$$= k(x_2 - x_1) + (\phi_{20} - \phi_{10}) \qquad (21.21)$$
$$= 2\pi\frac{\Delta x}{\lambda} + \Delta\phi_0$$

You can see that there are two contributions to the phase difference. $\Delta x = x_2 - x_1$, the distance between the two sources, is called **path-length difference**. It is the extra distance traveled by wave 2 on the way to the point where the two waves are combined. $\Delta\phi_0 = \phi_{20} - \phi_{10}$ is the *inherent phase difference* between the sources.

The condition of being in phase, where crests are aligned with crests and troughs with troughs, is $\Delta\phi = 0$, 2π, 4π, or any integer multiple of 2π rad. Thus the condition for maximum constructive interference is

Maximum constructive interference:
$$\Delta\phi = 2\pi\frac{\Delta x}{\lambda} + \Delta\phi_0 = m \cdot 2\pi \text{ rad} \qquad m = 0, 1, 2, 3, \ldots \qquad (21.22)$$

For identical sources, which have $\Delta\phi_0 = 0$ rad, maximum constructive interference occurs when $\Delta x = m\lambda$. That is, **two identical sources produce maximum constructive interference when the path-length difference is an integer number of wavelengths.**

FIGURE 21.21 Constructive and destructive interference of two waves traveling along the *x*-axis.

(a) Constructive interference

These two waves are in phase. Their crests are aligned.

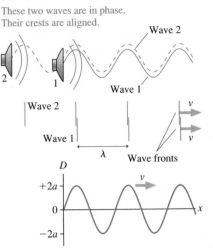

Their superposition produces a traveling wave moving to the right with amplitude 2a. This is maximum constructive interference.

(b) Destructive interference

These two waves are out of phase. The crests of one wave are aligned with the troughs of the other.

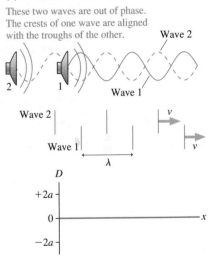

Their superposition produces a wave with zero amplitude. This is perfect destructive interference.

FIGURE 21.22 Two identical sources one wavelength apart.

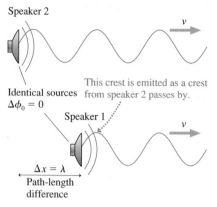

Speaker 2

This crest is emitted as a crest from speaker 2 passes by.

Identical sources
$\Delta\phi_0 = 0$

Speaker 1

$\Delta x = \lambda$
Path-length difference

The two waves are in phase ($\Delta\phi = 2\pi$ rad) and interfere constructively.

FIGURE 21.22 shows two identical sources (i.e., the two loudspeakers are doing the same thing at the same time), so $\Delta\phi_0 = 0$ rad. The path-length difference Δx is the extra distance traveled by the wave from loudspeaker 2 before it combines with loudspeaker 1. In this case, $\Delta x = \lambda$. Because a wave moves forward exactly one wavelength during one period, loudspeaker 1 emits a crest exactly as a crest of wave 2 passes by. The two waves are "in step," with $\Delta\phi = 2\pi$ rad, so the two waves interfere constructively to produce a wave of amplitude $2a$.

Perfect destructive interference, where the crests of one wave are aligned with the troughs of the other, occurs when two waves are *out of phase*, meaning that $\Delta\phi = \pi$, 3π, 5π, or any odd multiple of π rad. Thus the condition for perfect destructive interference is

> Perfect destructive interference:
>
> $$\Delta\phi = 2\pi\frac{\Delta x}{\lambda} + \Delta\phi_0 = \left(m + \frac{1}{2}\right)\cdot 2\pi \text{ rad} \qquad m = 0, 1, 2, 3, \ldots \qquad (21.23)$$

For identical sources, which have $\Delta\phi_0 = 0$ rad, perfect destructive interference occurs when $\Delta x = (m + \frac{1}{2})\lambda$. **That is, two identical sources produce perfect destructive interference when the path-length difference is a half-integer number of wavelengths.**

Two waves can be out of phase because the sources are located at different positions, because the sources themselves are out of phase, or because of a combination of these two. **FIGURE 21.23** illustrates these ideas by showing three different ways in which two waves interfere destructively. Each of these three arrangements creates waves with $\Delta\phi = \pi$ rad.

FIGURE 21.23 Destructive interference three ways.

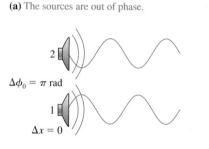

(a) The sources are out of phase.

2

$\Delta\phi_0 = \pi$ rad

1

$\Delta x = 0$

(b) Identical sources are separated by half a wavelength.

2

$\Delta\phi_0 = 0$ rad

1

$\Delta x = \frac{1}{2}\lambda$

(c) The sources are both separated and partially out of phase.

2

$\Delta\phi_0 = \frac{\pi}{2}$ rad

1

$\Delta x = \frac{1}{4}\lambda$

NOTE ▶ Don't confuse the phase difference of the waves ($\Delta\phi$) with the phase difference of the sources ($\Delta\phi_0$). It is $\Delta\phi$, the phase difference of the waves, that governs interference. ◀

EXAMPLE 21.7 **Interference between two sound waves**

You are standing in front of two side-by-side loudspeakers playing sounds of the same frequency. Initially there is almost no sound at all. Then one of the speakers is moved slowly away from you. The sound intensity increases as the separation between the speakers increases, reaching a maximum when the speakers are 0.75 m apart. Then, as the speaker continues to move, the intensity starts to decrease. What is the distance between the speakers when the sound intensity is again a minimum?

MODEL The changing sound intensity is due to the interference of two overlapped sound waves.

VISUALIZE Moving one speaker relative to the other changes the phase difference between the waves.

SOLVE A minimum sound intensity implies that the two sound waves are interfering destructively. Initially the loudspeakers are side by side, so the situation is as shown in Figure 21.23a with $\Delta x = 0$ and $\Delta\phi_0 = \pi$ rad. That is, the speakers themselves are out of phase. Moving one of the speakers does not change $\Delta\phi_0$, but it does change the path-length difference Δx and thus increases the overall phase difference $\Delta\phi$. Constructive interference, causing maximum intensity, is reached when

$$\Delta\phi = 2\pi\frac{\Delta x}{\lambda} + \Delta\phi_0 = 2\pi\frac{\Delta x}{\lambda} + \pi = 2\pi \text{ rad}$$

where we used $m = 1$ because this is the first separation giving constructive interference. The speaker separation at which this occurs is $\Delta x = \lambda/2$. This is the situation shown in FIGURE 21.24.

Because $\Delta x = 0.75$ m is $\lambda/2$, the sound's wavelength is $\lambda = 1.50$ m. The next point of destructive interference, with $m = 1$, occurs when

$$\Delta\phi = 2\pi\frac{\Delta x}{\lambda} + \Delta\phi_0 = 2\pi\frac{\Delta x}{\lambda} + \pi = 3\pi \text{ rad}$$

Thus the distance between the speakers when the sound intensity is again a minimum is

$$\Delta x = \lambda = 1.50 \text{ m}$$

ASSESS A separation of λ gives constructive interference for two *identical* speakers ($\Delta\phi_0 = 0$). Here the phase difference of π rad between the speakers (one is pushing forward as the other pulls back) gives destructive interference at this separation.

FIGURE 21.24 Two out-of-phase sources generate waves that are in phase if the sources are one half-wavelength apart.

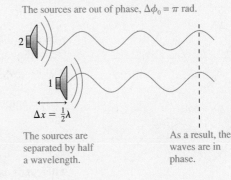

The sources are out of phase, $\Delta\phi_0 = \pi$ rad.

$\Delta x = \frac{1}{2}\lambda$

The sources are separated by half a wavelength.

As a result, the waves are in phase.

STOP TO THINK 21.4 Two loudspeakers emit waves with $\lambda = 2.0$ m. Speaker 2 is 1.0 m in front of speaker 1. What, if anything, can be done to cause constructive interference between the two waves?

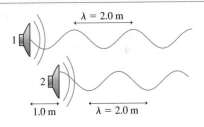

$\lambda = 2.0$ m

1.0 m $\lambda = 2.0$ m

a. Move speaker 1 forward (to the right) 1.0 m.
b. Move speaker 1 forward (to the right) 0.5 m.
c. Move speaker 1 backward (to the left) 0.5 m.
d. Move speaker 1 backward (to the left) 1.0 m.
e. Nothing. The situation shown already causes constructive interference.
f. Constructive interference is not possible for any placement of the speakers.

21.6 The Mathematics of Interference

Let's look more closely at the superposition of two waves. As two waves of equal amplitude and frequency travel together along the *x*-axis, the net displacement of the medium is

$$D = D_1 + D_2 = a\sin(kx_1 - \omega t + \phi_{10}) + a\sin(kx_2 - \omega t + \phi_{20})$$
$$= a\sin\phi_1 + a\sin\phi_2 \qquad (21.24)$$

where the phases ϕ_1 and ϕ_2 were defined in Equation 21.20.

A useful trigonometric identity is

$$\sin\alpha + \sin\beta = 2\cos\left[\tfrac{1}{2}(\alpha - \beta)\right]\sin\left[\tfrac{1}{2}(\alpha + \beta)\right] \qquad (21.25)$$

This identity is certainly not obvious, although it is easily proven by working backward from the right side. We can use this identity to write the net displacement of Equation 21.24 as

$$D = \left[2a\cos\left(\frac{\Delta\phi}{2}\right)\right]\sin(kx_{\text{avg}} - \omega t + (\phi_0)_{\text{avg}}) \qquad (21.26)$$

where $\Delta\phi = \phi_2 - \phi_1$ is the phase difference between the two waves, exactly as in Equation 21.21. $x_{\text{avg}} = (x_1 + x_2)/2$ is the average distance to the two sources and $(\phi_0)_{\text{avg}} = (\phi_{10} + \phi_{20})/2$ is the average phase constant of the sources.

The sine term shows that the superposition of the two waves is still a traveling wave. An observer would see a sinusoidal wave moving along the *x*-axis with the *same* wavelength and frequency as the original waves.

FIGURE 21.25 The interference of two waves for three different values of the phase difference.

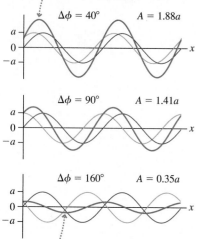

For $\Delta\phi = 40°$, the interference is constructive but not maximum constructive.

$\Delta\phi = 40°$ $A = 1.88a$

$\Delta\phi = 90°$ $A = 1.41a$

$\Delta\phi = 160°$ $A = 0.35a$

For $\Delta\phi = 160°$, the interference is destructive but not perfect destructive.

But how *big* is this wave compared to the two original waves? They each had amplitude a, but the amplitude of their superposition is

$$A = \left| 2a\cos\left(\frac{\Delta\phi}{2}\right) \right| \tag{21.27}$$

where we have used an absolute value sign because amplitudes must be positive. Depending upon the phase difference of the two waves, the amplitude of their superposition can be anywhere from zero (perfect destructive interference) to $2a$ (maximum constructive interference).

The amplitude has its maximum value $A = 2a$ if $\cos(\Delta\phi/2) = \pm 1$. This occurs when

$$\Delta\phi = m \cdot 2\pi \qquad \text{(maximum amplitude } A = 2a\text{)} \tag{21.28}$$

where m is an integer. Similarly, the amplitude is zero if $\cos(\Delta\phi/2) = 0$, which occurs when

$$\Delta\phi = \left(m + \tfrac{1}{2}\right) \cdot 2\pi \qquad \text{(minimum amplitude } A = 0\text{)} \tag{21.29}$$

Equations 21.28 and 21.29 are identical to the conditions of Equations 21.22 and 21.23 for constructive and destructive interference. We initially found these conditions by considering the alignment of the crests and troughs. Now we have confirmed them with an algebraic addition of the waves.

It is entirely possible, of course, that the two waves are neither exactly in phase nor exactly out of phase. Equation 21.27 allows us to calculate the amplitude of the superposition for any value of the phase difference. As an example, FIGURE 21.25 shows the calculated interference of two waves that differ in phase by 40°, by 90°, and by 160°.

EXAMPLE 21.8 **More interference of sound waves**

Two loudspeakers emit 500 Hz sound waves with an amplitude of 0.10 mm. Speaker 2 is 1.00 m behind speaker 1, and the phase difference between the speakers is 90°. What is the amplitude of the sound wave at a point 2.00 m in front of speaker 1?

MODEL The amplitude is determined by the interference of the two waves. Assume that the speed of sound has a room-temperature (20°C) value of 343 m/s.

SOLVE The amplitude of the sound wave is

$$A = |2a\cos(\Delta\phi/2)|$$

where $a = 0.10$ mm and the phase difference between the waves is

$$\Delta\phi = \phi_2 - \phi_1 = 2\pi\frac{\Delta x}{\lambda} + \Delta\phi_0$$

The sound's wavelength is

$$\lambda = \frac{v}{f} = \frac{343 \text{ m/s}}{500 \text{ Hz}} = 0.686 \text{ m}$$

Distances $x_1 = 2.00$ m and $x_2 = 3.00$ m are measured from the speakers, so the path-length difference is $\Delta x = 1.00$ m. We're given that the inherent phase difference between the speakers is $\Delta\phi_0 = \pi/2$ rad. Thus the phase difference at the observation point is

$$\Delta\phi = 2\pi\frac{\Delta x}{\lambda} + \Delta\phi_0 = 2\pi\frac{1.00 \text{ m}}{0.686 \text{ m}} + \frac{\pi}{2} \text{ rad} = 10.73 \text{ rad}$$

and the amplitude of the wave at this point is

$$A = \left| 2a\cos\left(\frac{\Delta\phi}{2}\right) \right| = \left| (0.200 \text{ mm})\cos\left(\frac{10.73}{2}\right) \right| = 0.121 \text{ mm}$$

ASSESS The interference is constructive because $A > a$, but less than maximum constructive interference.

Application: Thin-Film Optical Coatings

The shimmering colors of soap bubbles and oil slicks, as seen in the photo at the beginning of the chapter, are due to the interference of light waves. In fact, the idea of light-wave interference in one dimension has an important application in the optics industry, namely the use of **thin-film optical coatings**. These films, less than 1 μm (10^{-6} m) thick, are placed on glass surfaces, such as lenses, to control reflections from the glass. Antireflection coatings on the lenses in cameras, microscopes, and other optical equipment are examples of thin-film coatings.

FIGURE 21.26 shows a light wave of wavelength λ approaching a piece of glass that has been coated with a transparent film of thickness d whose index of refraction is n. The air-film boundary is a discontinuity at which the wave speed suddenly decreases, and you saw earlier, in Figure 21.8, that a discontinuity causes a reflection. Most of the light is transmitted into the film, but a little bit is reflected.

Furthermore, you saw in Figure 21.8 that the wave reflected from a discontinuity at which the speed decreases is *inverted* with respect to the incident wave. For a sinusoidal wave, which we're now assuming, the inversion is represented mathematically as a phase shift of π rad. The speed of a light wave decreases when it enters a material with a *larger* index of refraction. Thus **a light wave that reflects from a boundary at which the index of refraction increases has a phase shift of π rad.** There is no phase shift for the reflection from a boundary at which the index of refraction decreases. The reflection in Figure 21.26 is from a boundary between air ($n_{air} = 1.00$) and a transparent film with $n_{film} > n_{air}$, so the reflected wave is inverted due to the phase shift of π rad.

When the transmitted wave reaches the glass, most of it continues on into the glass but a portion is reflected back to the left. We'll assume that the index of refraction of the glass is larger than that of the film, $n_{glass} > n_{film}$, so this reflection also has a phase shift of π rad. This second reflection, after traveling back through the film, passes back into the air. There are now *two* equal-frequency waves traveling to the left, and these waves will interfere. If the two reflected waves are *in phase*, they will interfere constructively to cause a *strong reflection*. If the two reflected waves are *out of phase*, they will interfere destructively to cause a *weak reflection* or, if their amplitudes are equal, *no reflection* at all.

This suggests practical uses for thin-film optical coatings. The reflections from glass surfaces, even if weak, are often undesirable. For example, reflections degrade the performance of optical equipment. These reflections can be eliminated by coating the glass with a film whose thickness is chosen to cause *destructive* interference of the two reflected waves. This is an *antireflection coating*.

The amplitude of the reflected light depends on the phase difference between the two reflected waves. This phase difference is

$$\Delta\phi = \phi_2 - \phi_1 = (kx_2 + \phi_{20} + \pi \text{ rad}) - (kx_1 + \phi_{10} + \pi \text{ rad})$$
$$= 2\pi\frac{\Delta x}{\lambda_f} + \Delta\phi_0 \tag{21.30}$$

where we explicitly included the reflection phase shift of each wave. In this case, because *both* waves had a phase shift of π rad, the reflection phase shifts cancel.

The wavelength λ_f is the wavelength *in the film* because that's where the path-length difference Δx occurs. You learned in Chapter 20 that the wavelength in a transparent material with index of refraction n is $\lambda_f = \lambda/n$, where the unsubscripted λ is the wavelength in vacuum or air. That is, λ is the wavelength that we measure on "our" side of the air-film boundary.

The path-length difference between the two waves is $\Delta x = 2d$ because wave 2 travels through the film *twice* before rejoining wave 1. The two waves have a common origin—the initial division of the incident wave at the front surface of the film—so the inherent phase difference is $\Delta\phi_0 = 0$. Thus the phase difference of the two reflected waves is

$$\Delta\phi = 2\pi\frac{2d}{\lambda/n} = 2\pi\frac{2nd}{\lambda} \tag{21.31}$$

The interference is constructive, causing a strong reflection, when $\Delta\phi = m \cdot 2\pi$ rad. So when both reflected waves have a phase of π rad, constructive interference occurs for wavelengths

$$\lambda_C = \frac{2nd}{m} \qquad m = 1, 2, 3, \ldots \qquad \text{(constructive interference)} \tag{21.32}$$

FIGURE 21.26 The two reflections, one from the coating and one from the glass, interfere.

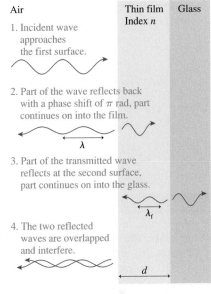

Air — Thin film Index n — Glass

1. Incident wave approaches the first surface.

2. Part of the wave reflects back with a phase shift of π rad, part continues on into the film.
λ

3. Part of the transmitted wave reflects at the second surface, part continues on into the glass.
λ_f

4. The two reflected waves are overlapped and interfere.

d

Antireflection coatings use the interference of light waves to nearly eliminate reflections from glass surfaces.

You will notice that m starts with 1, rather than 0, in order to give meaningful results. Destructive interference, with minimum reflection, requires $\Delta\phi = \left(m - \frac{1}{2}\right) \cdot 2\pi$ rad. This—again, when both waves have a phase shift of π rad—occurs for wavelengths

$$\lambda_{\mathrm{D}} = \frac{2nd}{m - \frac{1}{2}} \qquad m = 1, 2, 3, \ldots \qquad \text{(destructive interference)} \quad (21.33)$$

We've used $m - \frac{1}{2}$, rather than $m + \frac{1}{2}$, so that m can start with 1 to match the condition for constructive interference.

NOTE ▶ The exact condition for constructive or destructive interference is satisfied for only a few discrete wavelengths λ. Nonetheless, reflections are strongly enhanced (nearly constructive interference) for a range of wavelengths near λ_{C}. Likewise, there is a range of wavelengths near λ_{D} for which the reflection is nearly canceled. ◀

EXAMPLE 21.9 **Designing an antireflection coating**

Magnesium fluoride (MgF_2) is used as an antireflection coating on lenses. The index of refraction of MgF_2 is 1.39. What is the thinnest film of MgF_2 that works as an antireflection coating at $\lambda = 510$ nm, near the center of the visible spectrum?

MODEL Reflection is minimized if the two reflected waves interfere destructively.

SOLVE The film thicknesses that cause destructive interference at wavelength λ are

$$d = \left(m - \frac{1}{2}\right)\frac{\lambda}{2n}$$

The thinnest film has $m = 1$. Its thickness is

$$d = \frac{\lambda}{4n} = \frac{510 \text{ nm}}{4(1.39)} = 92 \text{ nm}$$

The film thickness is significantly less than the wavelength of visible light!

ASSESS The reflected light is completely eliminated (perfect destructive interference) only if the two reflected waves have equal amplitudes. In practice, they don't. Nonetheless, the reflection is reduced from $\approx 4\%$ of the incident intensity for "bare glass" to well under 1%. Furthermore, the intensity of reflected light is much reduced across most of the visible spectrum (400–700 nm), even though the phase difference deviates more and more from π rad as the wavelength moves away from 510 nm. It is the increasing reflection at the ends of the visible spectrum ($\lambda \approx 400$ nm and $\lambda \approx 700$ nm), where $\Delta\phi$ deviates significantly from π rad, that gives a reddish-purple tinge to the lenses on cameras and binoculars. Homework problems will let you explore situations where only one of the two reflections has a reflection phase shift of π rad.

21.7 Interference in Two and Three Dimensions

Ripples on a lake move in two dimensions. The glow from a lightbulb spreads outward as a spherical wave. A circular or spherical wave can be written

$$D(r, t) = a\sin(kr - \omega t + \phi_0) \qquad (21.34)$$

where r is the distance measured outward from the source. Equation 21.34 is our familiar wave equation with the one-dimensional coordinate x replaced by a more general radial coordinate r. Strictly speaking, the amplitude a of a circular or spherical wave diminishes as r increases. However, we will assume that a remains essentially constant over the region in which we study the wave. FIGURE 21.27 shows the wave-front diagram for a circular or spherical wave. Recall that the wave fronts represent the *crests* of the wave and are spaced by the wavelength λ.

What happens when two circular or spherical waves overlap? For example, imagine two paddles oscillating up and down on the surface of a pond. We will assume that the two paddles oscillate with the same frequency and amplitude and that they are in phase. FIGURE 21.28 shows the wave fronts of the two waves. The ripples overlap as they travel, and, as was the case in one dimension, this causes interference.

Constructive interference with $A = 2a$ occurs where two crests align or two troughs align. Several locations of constructive interference are marked in Figure 21.28. Intersecting wave fronts are points where two crests are aligned. It's a bit harder to

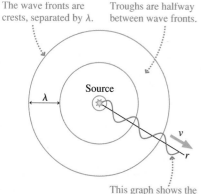

FIGURE 21.27 A circular or spherical wave.

The wave fronts are crests, separated by λ.

Troughs are halfway between wave fronts.

λ

Source

v

r

This graph shows the displacement of the medium.

visualize, but two troughs are aligned when a midpoint between two wave fronts is overlapped with another midpoint between two wave fronts. Destructive interference with $A = 0$ occurs where the crest of one wave aligns with a trough of the other wave. Several points of destructive interference are also indicated in Figure 21.28.

A picture on a page is static, but **the wave fronts are in motion.** Try to imagine the wave fronts of Figure 21.28 expanding outward as new circular rings are born at the sources. The waves will move forward half a wavelength during half a period, causing the crests in Figure 21.28 to be replaced by troughs while the troughs become crests.

The important point to recognize is that **the motion of the waves does not affect the points of constructive and destructive interference.** Points in the figure where two crests overlap will become points where two troughs overlap, but this overlap is still constructive interference. Similarly, points in the figure where a crest and a trough overlap will become a point where a trough and a crest overlap—still destructive interference.

The mathematical description of interference in two or three dimensions is very similar to that of one-dimensional interference. The net displacement of a particle in the medium is

$$D = D_1 + D_2 = a\sin(kr_1 - \omega t + \phi_{10}) + a\sin(kr_2 - \omega t + \phi_{20}) \quad (21.35)$$

The only difference between Equation 21.35 and the earlier one-dimensional Equation 21.24 is that the linear coordinates x_1 and x_2 have been changed to radial coordinates r_1 and r_2. Thus our conclusions are unchanged. The superposition of the two waves yields a wave traveling outward with amplitude

$$A = \left| 2a\cos\left(\frac{\Delta\phi}{2}\right) \right| \quad (21.36)$$

where the phase difference, with x replaced by r, is now

$$\Delta\phi = 2\pi\frac{\Delta r}{\lambda} + \Delta\phi_0 \quad (21.37)$$

The term $2\pi(\Delta r/\lambda)$ is the phase difference that arises when the waves travel different distances from the sources to the point at which they combine. Δr itself is the *path-length difference.* As before, $\Delta\phi_0$ is any inherent phase difference of the sources themselves.

Maximum constructive interference with $A = 2a$ occurs, just as in one dimension, at those points where $\cos(\Delta\phi/2) = \pm 1$. Similarly, perfect destructive interference occurs at points where $\cos(\Delta\phi/2) = 0$. The conditions for constructive and destructive interference are

Maximum constructive interference:

$$\Delta\phi = 2\pi\frac{\Delta r}{\lambda} + \Delta\phi_0 = m \cdot 2\pi$$

$$m = 0, 1, 2, \ldots \quad (21.38)$$

Perfect destructive interference:

$$\Delta\phi = 2\pi\frac{\Delta r}{\lambda} + \Delta\phi_0 = \left(m + \frac{1}{2}\right) \cdot 2\pi$$

For two identical sources (i.e., sources that oscillate in phase with $\Delta\phi_0 = 0$), the conditions for constructive and destructive interference are simple:

Constructive: $\quad \Delta r = m\lambda$

Destructive: $\quad \Delta r = \left(m + \frac{1}{2}\right)\lambda$ $\qquad$ (identical sources) $\quad$ (21.39)

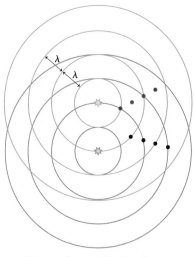

FIGURE 21.28 The overlapping ripple patterns of two sources. Several points of constructive and destructive interference are noted.

Two in-phase sources emit circular or spherical waves.

- Points of constructive interference. A crest is aligned with a crest, or a trough with a trough.

- Points of destructive interference. A crest is aligned with a trough of another wave.

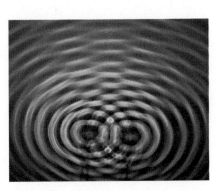

Two overlapping water waves create an interference pattern.

FIGURE 21.29 The path-length difference Δr determines whether the interference at a particular point is constructive or destructive.

• At A, $\Delta r_A = \lambda$, so this is a point of constructive interference.

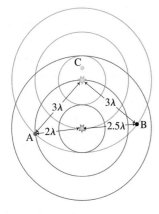

• At B, $\Delta r_B = \frac{1}{2}\lambda$, so this is a point of destructive interference.

The waves from two identical sources interfere constructively at points where the path-length difference is an integer number of wavelengths because, for these values of Δr, crests are aligned with crests and troughs with troughs. **The waves interfere destructively at points where the path-length difference is a half-integer number of wavelengths** because, for these values of Δr, crests are aligned with troughs. These two statements are the essence of interference.

NOTE ▶ Equation 21.39 applies only if the sources are in phase. If the sources are not in phase, you must use the more general Equation 21.38 to locate the points of constructive and destructive interference. ◀

Wave fronts are spaced exactly one wavelength apart; hence we can measure the distances r_1 and r_2 simply by counting the rings in the wave-front pattern. In FIGURE 21.29, which is based on Figure 21.28, point A is distance $r_1 = 3\lambda$ from the first source and $r_2 = 2\lambda$ from the second. The path-length difference is $\Delta r_A = 1\lambda$, the condition for the maximum constructive interference of identical sources. Point B has $\Delta r_B = \frac{1}{2}\lambda$, so it is a point of perfect destructive interference.

NOTE ▶ Interference is determined by Δr, the path-length *difference,* rather than by r_1 or r_2. ◀

STOP TO THINK 21.5 The interference at point C in Figure 21.29 is

a. Maximum constructive.
c. Perfect destructive.
e. There is no interference at point C.

b. Constructive, but less than maximum.
d. Destructive, but not perfect.

We can now locate the points of maximum constructive interference, for which $\Delta r = m\lambda$, by drawing a line through *all* the points at which $\Delta r = 0$, another line through all the points at which $\Delta r = \lambda$, and so on. These lines, shown in red in FIGURE 21.30, are called **antinodal lines.** They are analogous to the antinodes of a standing wave, hence the name. An antinode is a *point* of maximum constructive interference; for circular waves, oscillation at maximum amplitude occurs along a continuous *line.* Similarly, destructive interference occurs along lines called **nodal lines.** The displacement is *always zero* along these lines, just as it is at a node in a standing-wave pattern.

FIGURE 21.30 The points of constructive and destructive interference fall along antinodal and nodal lines.

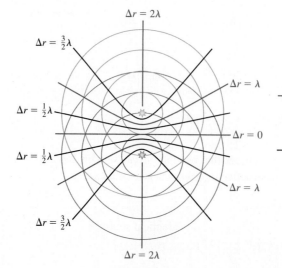

— Antinodal lines, constructive interference, oscillation with maximum amplitude. Intensity is at its maximum value.

— Nodal lines, destructive interference, no oscillation. Intensity is zero.

A Problem-Solving Strategy for Interference Problems

The information in this section is the basis of a strategy for solving interference problems. This strategy applies equally well to interference in one dimension if you use Δx instead of Δr.

PROBLEM-SOLVING STRATEGY 21.1 ## Interference of two waves (MP)

MODEL Make simplifying assumptions, such as assuming waves are circular and of equal amplitude.

VISUALIZE Draw a picture showing the sources of the waves and the point where the waves interfere. Give relevant dimensions. Identify the distances r_1 and r_2 from the sources to the point. Note any phase difference $\Delta\phi_0$ between the two sources.

SOLVE The interference depends on the path-length difference $\Delta r = r_2 - r_1$ and the source phase difference $\Delta\phi_0$.

Constructive: $\quad \Delta\phi = 2\pi \dfrac{\Delta r}{\lambda} + \Delta\phi_0 = m \cdot 2\pi$

$$m = 0, 1, 2, \ldots$$

Destructive: $\quad \Delta\phi = 2\pi \dfrac{\Delta r}{\lambda} + \Delta\phi_0 = \left(m + \dfrac{1}{2}\right) \cdot 2\pi$

For identical sources ($\Delta\phi_0 = 0$), the interference is maximum constructive if $\Delta r = m\lambda$, perfect destructive if $\Delta r = \left(m + \frac{1}{2}\right)\lambda$.

ASSESS Check that your result has the correct units, is reasonable, and answers the question.

Exercise 18 /

EXAMPLE 21.10 **Two-dimensional interference between two loudspeakers**

Two loudspeakers in a plane are 2.0 m apart and in phase with each other. Both emit 700 Hz sound waves into a room where the speed of sound is 341 m/s. A listener stands 5.0 m in front of the loudspeakers and 2.0 m to one side of the center. Is the interference at this point maximum constructive, perfect destructive, or in between? How will the situation differ if the loudspeakers are out of phase?

MODEL The two speakers are sources of in-phase, spherical waves. The overlap of these waves causes interference.

VISUALIZE FIGURE 21.31 shows the loudspeakers and defines the distances r_1 and r_2 to the point of observation. The figure includes dimensions and notes that $\Delta\phi_0 = 0$ rad.

FIGURE 21.31 Pictorial representation of the interference between two loudspeakers.

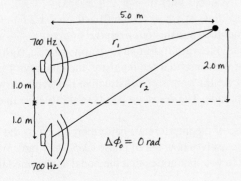

SOLVE It's not r_1 and r_2 that matter, but the *difference* Δr between them. From the geometry of the figure we can calculate that

$$r_1 = \sqrt{(5.0 \text{ m})^2 + (1.0 \text{ m})^2} = 5.10 \text{ m}$$
$$r_2 = \sqrt{(5.0 \text{ m})^2 + (3.0 \text{ m})^2} = 5.83 \text{ m}$$

Thus the path-length difference is $\Delta r = r_2 - r_1 = 0.73$ m. The wavelength of the sound waves is

$$\lambda = \frac{v}{f} = \frac{341 \text{ m/s}}{700 \text{ Hz}} = 0.487 \text{ m}$$

In terms of wavelengths, the path-length difference is $\Delta r/\lambda = 1.50$, or

$$\Delta r = \frac{3}{2}\lambda$$

Because the sources are in phase ($\Delta\phi_0 = 0$), this is the condition for *destructive* interference. If the sources were out of phase ($\Delta\phi_0 = \pi$ rad), then the phase difference of the waves at the listener would be

$$\Delta\phi = 2\pi \frac{\Delta r}{\lambda} + \Delta\phi_0 = 2\pi\left(\frac{3}{2}\right) + \pi \text{ rad} = 4\pi \text{ rad}$$

This is an integer multiple of 2π rad, so in this case the interference would be *constructive*.

ASSESS Both the path-length difference *and* any inherent phase difference of the sources must be considered when evaluating interference.

Picturing Interference

A *contour map* is a useful way to visualize an interference pattern. FIGURE 21.32a shows the superposition of the waves from two identical sources ($\Delta\phi_0 = 0$) emitting waves with $\lambda = 1$ m. The sources, indicated with black dots, are located two wavelengths apart at $y = \pm 1$ m. Positive displacements are shown in red, with the deepest red representing the maximum displacement of the wave at this instant in time. These are the points where the crests of the individual waves interfere constructively to give $D = 2a$. Negative displacements are blue, with the darkest blue being the most negative displacement of the wave. These are also points of constructive interference, with two troughs overlapping to give $D = -2a$.

FIGURE 21.32 A contour map of the interference pattern of two sources. The graph on the right side of each figure shows the wave intensity along a vertical line at $x = 4$ m.

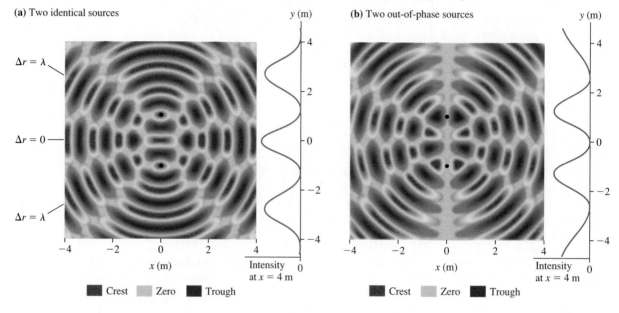

(a) Two identical sources

(b) Two out-of-phase sources

To understand this figure, try to visualize the waves expanding outward from the center. The red-blue-red-blue-red-··· pattern of crests and troughs moves outward along the antinodal lines as a *traveling wave* of amplitude $A = 2a$. Nothing ever happens along the nodal lines, where the amplitude is always zero.

Suppose you were to observe the *intensity* of the wave as it crosses the vertical line at $x = 4$ m on the right edge of the figure. If, for example, these are sound waves, you could listen to (or measure with a microphone) the sound intensity as you walk from $(x, y) = (4\text{ m}, -4\text{ m})$ at the bottom of the figure to $(x, y) = (4\text{ m}, 4\text{ m})$ at the top. The intensity is zero as you cross the nodal lines at $y \approx \pm 1$ m $\left(\Delta r = \frac{1}{2}\lambda\right)$. The intensity is maximum at the antinodal lines at $y = 0$ ($\Delta r = 0$) and $y \approx \pm 2.5$ m ($\Delta r = \lambda$), where a wave of maximum amplitude streams out from the sources.

The intensity is shown in the rather unusual graph on the right side of Figure 21.32a. It is unusual in the sense that the intensity, the quantity of interest, is graphed to the left. The peaks are the points of constructive interference, where you would measure maximum amplitude. The zeros are points of destructive interference, where the intensity is zero.

FIGURE 21.32b is a contour map of the interference pattern produced by the same two sources but with the sources themselves now out of phase ($\Delta\phi_0 = \pi$ rad). We'll leave the investigation of this figure to you, but notice that the nodal and antinodal lines are reversed from those of Figure 21.32a.

EXAMPLE 21.11 **The intensity of two interfering loudspeakers**

Two loudspeakers in a plane are 6.0 m apart and in phase. They emit equal-amplitude sound waves with a wavelength of 1.0 m. Each speaker alone creates sound with intensity I_0. An observer at point A is 10 m in front of the plane containing the two loudspeakers and centered between them. A second observer at point B is 10 m directly in front of one of the speakers. In terms of I_0, what are the intensity I_A at point A and the intensity I_B at point B?

MODEL The two speakers are sources of in-phase waves. The overlap of these waves causes interference.

VISUALIZE **FIGURE 21.33** shows the two loudspeakers and the two points of observation. Distances r_1 and r_2 are defined for point B.

FIGURE 21.33 Pictorial representation of the interference between two loudspeakers.

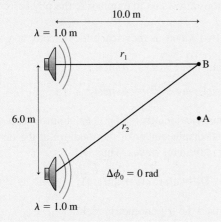

SOLVE Let the amplitude of the wave from each speaker be a. The intensity of a wave is proportional to the square of the amplitude,

so the intensity of each speaker alone is $I_0 = ca^2$, where c is an unknown proportionality constant. Point A is a point of constructive interference because the speakers are in phase ($\Delta\phi_0 = 0$) and the path-length difference is $\Delta r = 0$. The amplitude at this point is given by Equation 21.36:

$$A_A = \left| 2a\cos\left(\frac{\Delta\phi}{2}\right) \right| = 2a\cos(0) = 2a$$

Consequently, the intensity at this point is

$$I_A = cA_A^2 = c(2a)^2 = 4ca^2 = 4I_0$$

The intensity at A is four times that of either speaker played alone.

At point B, the path-length difference is

$$\Delta r = \sqrt{(10.0\text{ m})^2 + (6.0\text{ m})^2} - 10.0\text{ m} = 1.662\text{ m}$$

The phase difference of the waves at this point is

$$\Delta\phi = 2\pi\frac{\Delta r}{\lambda} = 2\pi\frac{1.662\text{ m}}{1.0\text{ m}} = 10.44\text{ rad}$$

Consequently, the amplitude at B is

$$A_B = \left| 2a\cos\left(\frac{\Delta\phi}{2}\right) \right| = |2a\cos(5.22\text{ rad})| = 0.972a$$

Thus the intensity at this point is

$$I_B = cA_B^2 = c(0.972a)^2 = 0.95ca^2 = 0.95I_0$$

ASSESS Although B is directly in front of one of the speakers, superposition of the two waves results in an intensity that is less than it would be if this speaker played alone.

STOP TO THINK 21.6 These two loudspeakers are in phase. They emit equal-amplitude sound waves with a wavelength of 1.0 m. At the point indicated, is the interference maximum constructive, perfect destructive, or something in between?

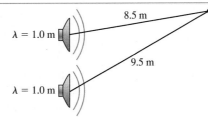

21.8 Beats

Thus far we have looked at the superposition of sources having the same wavelength and frequency. We can also use the principle of superposition to investigate a phenomenon that is easily demonstrated with two sources of slightly different frequency.

If you listen to two sounds with very different frequencies, such as a high note and a low note, you hear two distinct tones. But if the frequency difference is very small, just one or two hertz, then you hear a single tone whose intensity is *modulated* once or twice every second. That is, the sound goes up and down in volume, loud, soft, loud, soft,..., making a distinctive sound pattern called **beats**.

Consider two sinusoidal waves traveling along the x-axis with angular frequencies $\omega_1 = 2\pi f_1$ and $\omega_2 = 2\pi f_2$. The two waves are

$$D_1 = a\sin(k_1 x - \omega_1 t + \phi_{10})$$
$$D_2 = a\sin(k_2 x - \omega_2 t + \phi_{20})$$

(21.40)

where the subscripts 1 and 2 indicate that the frequencies, wave numbers, and phase constants of the two waves may be different.

To simplify the analysis, let's make several assumptions:

1. The two waves have the same amplitude a,
2. A detector, such as your ear, is located at the origin ($x = 0$),
3. The two sources are in phase ($\phi_{10} = \phi_{20}$), and
4. The source phases happen to be $\phi_{10} = \phi_{20} = \pi$ rad.

None of these assumptions is essential to the outcome. All could be otherwise and we would still come to basically the same conclusion, but the mathematics would be far messier. Making these assumptions allows us to emphasize the physics with the least amount of mathematics.

With these assumptions, the two waves as they reach the detector at $x = 0$ are

$$D_1 = a\sin(-\omega_1 t + \pi) = a\sin\omega_1 t$$
$$D_2 = a\sin(-\omega_2 t + \pi) = a\sin\omega_2 t$$

(21.41)

where we've used the trigonometric identity $\sin(\pi - \theta) = \sin\theta$. The principle of superposition tells us that the *net* displacement of the medium at the detector is the sum of the displacements of the individual waves. Thus

$$D = D_1 + D_2 = a(\sin\omega_1 t + \sin\omega_2 t)$$

(21.42)

Earlier, for interference, we used the trigonometric identity

$$\sin\alpha + \sin\beta = 2\cos\left[\tfrac{1}{2}(\alpha - \beta)\right]\sin\left[\tfrac{1}{2}(\alpha + \beta)\right]$$

We can use this identity again to write Equation 21.42 as

$$D = 2a\cos\left[\tfrac{1}{2}(\omega_1 - \omega_2)t\right]\sin\left[\tfrac{1}{2}(\omega_1 + \omega_2)t\right]$$
$$= \left[2a\cos(\omega_{\text{mod}}t)\right]\sin(\omega_{\text{avg}}t)$$

(21.43)

where $\omega_{\text{avg}} = \tfrac{1}{2}(\omega_1 + \omega_2)$ is the *average* angular frequency and $\omega_{\text{mod}} = \tfrac{1}{2}(\omega_1 - \omega_2)$ is called the *modulation frequency*.

We are interested in the situation when the two frequencies are very nearly equal: $\omega_1 \approx \omega_2$. In that case, ω_{avg} hardly differs from either ω_1 or ω_2 while ω_{mod} is very near to—but not exactly—zero. When ω_{mod} is very small, the term $\cos(\omega_{\text{mod}}t)$ oscillates *very* slowly. We have grouped it with the $2a$ term because, together, they provide a slowly changing "amplitude" for the rapid oscillation at frequency ω_{avg}.

FIGURE 21.34 is a history graph of the wave at the detector ($x = 0$). It shows the oscillation of the air against your eardrum at frequency $f_{\text{avg}} = \omega_{\text{avg}}/2\pi = \tfrac{1}{2}(f_1 + f_2)$. This oscillation determines the note you hear; it differs little from the two notes at frequencies f_1 and f_2. We are especially interested in the time-dependent amplitude, shown as a dashed line, that is given by the term $2a\cos(\omega_{\text{mod}}t)$. This periodically varying amplitude is called a **modulation** of the wave, which is where ω_{mod} gets its name.

As the amplitude rises and falls, the sound alternates as loud, soft, loud, soft, and so on. But that is exactly what you hear when you listen to beats! The alternating loud and soft sounds arise from the two waves being alternately in phase and out of phase, causing constructive and then destructive interference.

FIGURE 21.34 Beats are caused by the superposition of two waves of nearly identical frequency.

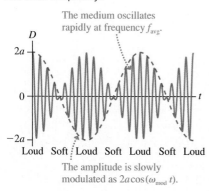

The medium oscillates rapidly at frequency f_{avg}.

Loud Soft Loud Soft Loud Soft Loud

The amplitude is slowly modulated as $2a\cos(\omega_{\text{mod}}t)$.

Imagine two people walking side by side at just slightly different paces. Initially both of their right feet hit the ground together, but after a while they get out of step. A little bit later they are back in step and the process alternates. The sound waves are doing the same. Initially the crests of each wave, of amplitude a, arrive together at your ear and the net displacement is doubled to $2a$. But after a while the two waves, being of slightly different frequency, get out of step and a crest of one arrives with a trough of the other. When this happens, the two waves cancel each other to give a net displacement of zero. This process alternates over and over, loud and soft.

Notice, from the figure, that the sound intensity rises and falls *twice* during one cycle of the modulation envelope. Each "loud-soft-loud" is one beat, so the **beat frequency** f_{beat}, which is the number of beats per second, is *twice* the modulation frequency $f_{mod} = \omega_{mod}/2\pi$. From the above definition of ω_{mod}, the beat frequency is

$$f_{beat} = 2f_{mod} = 2\frac{\omega_{mod}}{2\pi} = 2 \cdot \frac{1}{2}\left(\frac{\omega_1}{2\pi} - \frac{\omega_2}{2\pi}\right) = f_1 - f_2 \qquad (21.44)$$

where, to keep f_{beat} from being negative, we will always let f_1 be the larger of the two frequencies. The beat frequency is simply the *difference* between the two individual frequencies.

EXAMPLE 21.12 **Detecting bats with beats**

The little brown bat is a common species in North America. It emits echolocation pulses at a frequency of 40 kHz, well above the range of human hearing. To allow researchers to "hear" these bats, the bat detector shown in **FIGURE 21.35** combines the bat's sound wave at frequency f_1 with a wave of frequency f_2 from a tunable oscillator. The resulting beat frequency is then amplified and sent to a loudspeaker. To what frequency should the tunable oscillator be set to produce an audible beat frequency of 3 kHz?

SOLVE Combining two waves with different frequencies gives a beat frequency

$$f_{beat} = f_1 - f_2$$

A beat frequency will be generated at 3 kHz if the oscillator frequency and the bat frequency *differ* by 3 kHz. An oscillator frequency of either 37 kHz or 43 kHz will work nicely.

ASSESS The electronic circuitry of radios, televisions, and cell phones makes extensive use of *mixers* to generate difference frequencies.

FIGURE 21.35 The operation of a bat detector.

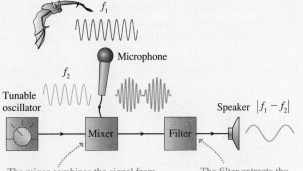

The mixer combines the signal from the bat with a sinusoidal wave from an oscillator. The result is a modulated wave.

The filter extracts the beat frequency, which is sent to the speaker.

Beats aren't limited to sound waves. **FIGURE 21.36** shows a graphical example of beats. Two "fences" of slightly different frequencies are superimposed on each other. The difference in the two frequencies is two lines per inch. You can confirm, with a ruler, that the figure has two "beats" per inch, in agreement with Equation 21.44.

Beats are important in many other situations. For example, you have probably seen movies where rotating wheels seem to turn slowly backward. Why is this? Suppose the movie camera is shooting at 30 frames per second but the wheel is rotating 32 times per second. The combination of the two produces a "beat" of 2 Hz, meaning that the wheel appears to rotate only twice per second. The same is true if the wheel is rotating 28 times per second, but in this case, where the wheel frequency slightly lags the camera frequency, it appears to rotate *backward* twice per second!

FIGURE 21.36 A graphical example of beats.

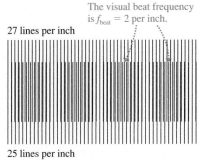

You hear three beats per second when two sound tones are generated. The frequency of one tone is 610 Hz. The frequency of the other is

a. 604 Hz
d. 616 Hz

b. 607 Hz
e. Either a or d.

c. 613 Hz
f. Either b or c.

CHALLENGE EXAMPLE 21.13 | **An airplane landing system**

Your firm has been hired to design a system that allows airplane pilots to make instrument landings in rain or fog. You've decided to place two radio transmitters 50 m apart on either side of the runway. These two transmitters will broadcast the same frequency, but out of phase with each other. This will cause a nodal line to extend straight off the end of the runway. As long as the airplane's receiver is silent, the pilot knows she's directly in line with the runway. If she drifts to one side or the other, the radio will pick up a signal and sound a warning beep. To have sufficient accuracy, the first intensity maxima need to be 60 m on either side of the nodal line at a distance of 3.0 km. What frequency should you specify for the transmitters?

MODEL The two transmitters are sources of out-of-phase, circular waves. The overlap of these waves produces an interference pattern.

VISUALIZE For out-of-phase sources, the center line—with zero path-length difference—is a nodal line of perfect destructive interference because the two signals always arrive out of phase. FIGURE 21.37 shows the nodal line, extending straight off the runway, and the first antinodal line—the points of maximum con-

structive interference—on either side. Comparing this to Figure 21.30, where the two sources were in phase, you can see that the nodal and antinodal lines have been reversed.

SOLVE Point P, 60 m to the side at a distance of 3000 m, needs to be a point of maximum constructive interference. The distances are

$$r_1 = \sqrt{(3000 \text{ m})^2 + (60 \text{ m} - 25 \text{ m})^2} = 3000.204 \ m$$
$$r_2 = \sqrt{(3000 \text{ m})^2 + (60 \text{ m} + 25 \text{ m})^2} = 3001.204 \ m$$

We needed to keep several extra significant figures because we're looking for the difference between two numbers that are almost the same. The path-length difference at P is

$$\Delta r = r_2 - r_1 = 1.000 \text{ m}$$

We know, for out-of-phase transmitters, that the phase difference of the sources is $\Delta\phi_0 = \pi$ rad. The first maximum will occur where the phase difference between the waves is $\Delta\phi = 1 \cdot 2\pi$ rad. Thus the condition that we must satisfy at P is

$$\Delta\phi = 2\pi \text{ rad} = 2\pi \frac{\Delta r}{\lambda} + \pi \text{ rad}$$

Solving for λ, we find

$$\lambda = 2 \Delta r = 2.00 \text{ m}$$

Consequently, the required frequency is

$$f = \frac{c}{\lambda} = \frac{3.00 \times 10^8 \text{ m/s}}{2.00 \text{ m}} = 1.50 \times 10^8 \text{ Hz} = 150 \text{ MHz}$$

ASSESS 150 MHz is slightly higher than the frequencies of FM radio ($\approx$ 100 MHz) but is well within the radio frequency range. Notice that the condition to be satisfied at P is that the path-length difference must be $\frac{1}{2}\lambda$. This makes sense. A path-length difference of $\frac{1}{2}\lambda$ contributes π rad to the phase difference. When combined with the π rad from the out-of-phase sources, the total phase difference of 2π rad creates constructive interference.

FIGURE 21.37 Pictorial representation of the landing system.

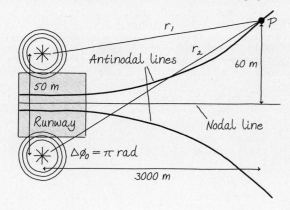

SUMMARY

The goal of Chapter 21 has been to understand and use the idea of superposition.

General Principles

Principle of Superposition

The displacement of a medium when more than one wave is present is the sum of the displacements due to each individual wave.

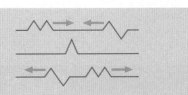

Important Concepts

Standing waves are due to the superposition of two traveling waves moving in opposite directions.

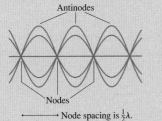

Antinodes

Nodes

Node spacing is $\frac{1}{2}\lambda$.

The amplitude at position x is

$$A(x) = 2a \sin kx$$

where a is the amplitude of each wave.

The boundary conditions determine which standing-wave frequencies and wavelengths are allowed. The allowed standing waves are **modes** of the system.

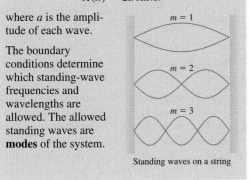

$m = 1$

$m = 2$

$m = 3$

Standing waves on a string

Interference

In general, the superposition of two or more waves into a single wave is called interference.

Maximum constructive interference occurs where crests are aligned with crests and troughs with troughs. These waves are in phase. The maximum displacement is $A = 2a$.

Perfect destructive interference occurs where crests are aligned with troughs. These waves are out of phase. The amplitude is $A = 0$.

Interference depends on the phase difference $\Delta\phi$ between the two waves.

Constructive: $\Delta\phi = 2\pi \dfrac{\Delta r}{\lambda} + \Delta\phi_0 = m \cdot 2\pi$

Destructive: $\Delta\phi = 2\pi \dfrac{\Delta r}{\lambda} + \Delta\phi_0 = \left(m + \dfrac{1}{2}\right) \cdot 2\pi$

Δr is the path-length difference of the two waves, and $\Delta\phi_0$ is any phase difference between the sources. For identical sources (in phase, $\Delta\phi_0 = 0$):

Interference is constructive if the path-length difference $\Delta r = m\lambda$.

Interference is destructive if the path-length difference $\Delta r = \left(m + \frac{1}{2}\right)\lambda$.

The amplitude at a point where the phase difference is $\Delta\phi$ is $A = \left| 2a \cos\left(\dfrac{\Delta\phi}{2}\right) \right|$.

Antinodal lines, constructive interference. $A = 2a$

Nodal lines, destructive interference. $A = 0$

Applications

Boundary conditions

Strings, electromagnetic waves, and sound waves in closed-closed tubes must have nodes at both ends:

$$\lambda_m = \frac{2L}{m} \qquad f_m = m\frac{v}{2L} = mf_1$$

where $m = 1, 2, 3, \ldots$.

The frequencies and wavelengths are the same for a sound wave in an open-open tube, which has antinodes at both ends.

A sound wave in an open-closed tube must have a node at the closed end but an antinode at the open end. This leads to

$$\lambda_m = \frac{4L}{m} \qquad f_m = m\frac{v}{4L} = mf_1$$

where $m = 1, 3, 5, 7, \ldots$.

Beats (loud-soft-loud-soft modulations of intensity) occur when two waves of slightly different frequency are superimposed.

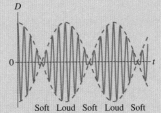

Soft Loud Soft Loud Soft

The beat frequency between waves of frequencies f_1 and f_2 is

$$f_{\text{beat}} = f_1 - f_2$$

Terms and Notation

principle of superposition	mode	path-length difference, Δx or Δr
standing wave	interference	thin-film optical coating
node	in phase	antinodal line
antinode	constructive interference	nodal line
amplitude function, $A(x)$	out of phase	beats
boundary condition	destructive interference	modulation
fundamental frequency, f_1	phase difference, $\Delta\phi$	beat frequency, f_{beat}
harmonic		

CONCEPTUAL QUESTIONS

1. **FIGURE Q21.1** shows a standing wave oscillating on a string at frequency f_0.

   FIGURE Q21.1

   a. What mode (m-value) is this?
   b. How many antinodes will there be if the frequency is doubled to $2f_0$?

2. If you take snapshots of a standing wave on a string, there are certain instants when the string is totally flat. What has happened to the energy of the wave at those instants?

3. **FIGURE Q21.3** shows the displacement of a standing sound wave in a 32-cm-long horizontal tube of air open at both ends.

   a. What mode (m-value) is this?
   b. Are the air molecules moving horizontally or vertically? Explain.
   c. At what distances from the left end of the tube do the molecules oscillate with maximum amplitude?
   d. At what distances from the left end of the tube does the air pressure oscillate with maximum amplitude?

4. An organ pipe is tuned to exactly 384 Hz when the room temperature is 20°C. If the room temperature later increases to 22°C, does the pipe's frequency increase, decrease, or stay the same? Explain.

5. If you pour liquid into a tall, narrow glass, you may hear sound with a steadily rising pitch. What is the source of the sound? And why does the pitch rise as the glass fills?

6. A flute filled with helium will, until the helium escapes, play notes at a much higher pitch than normal. Why?

7. In music, two notes are said to be an *octave* apart when one note is exactly twice the frequency of the other. Suppose you have a guitar string playing frequency f_0. To increase the frequency by an octave, to $2f_0$, by what factor would you have to (a) increase the tension or (b) decrease the length?

8. **FIGURE Q21.8** is a snapshot graph of two plane waves passing through a region of space. Each wave has a 2.0 mm amplitude and the same wavelength. What is the net displacement of the medium at points a, b, and c?

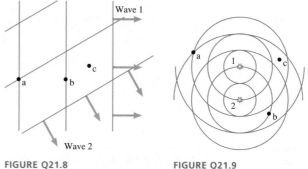

FIGURE Q21.8   **FIGURE Q21.9**

9. **FIGURE Q21.9** shows the circular waves emitted by two in-phase sources. Are points a, b, and c points of maximum constructive interference or perfect destructive interference? Explain.

10. A trumpet player hears 5 beats per second when she plays a note and simultaneously sounds a 440 Hz tuning fork. After pulling her tuning valve out to slightly increase the length of her trumpet, she hears 3 beats per second against the tuning fork. Was her initial frequency 435 Hz or 445 Hz? Explain.

For Q21.3:

D

FIGURE Q21.3

EXERCISES AND PROBLEMS

Problems labeled [   ] integrate material from earlier chapters.

Exercises

Section 21.1 The Principle of Superposition

1. | **FIGURE EX21.1** is a snapshot graph at $t = 0$ s of two waves approaching each other at 1.0 m/s. Draw six snapshot graphs, stacked vertically, showing the string at 1 s intervals from $t = 1$ s to $t = 6$ s.

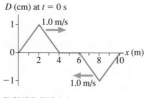

FIGURE EX21.1

2. | **FIGURE EX21.2** is a snapshot graph at $t = 0$ s of two waves approaching each other at 1.0 m/s. Draw six snapshot graphs, stacked vertically, showing the string at 1 s intervals from $t = 1$ s to $t = 6$ s.

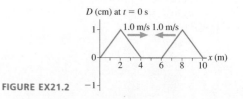

FIGURE EX21.2

3. ‖ **FIGURE EX21.3** is a snapshot graph at $t = 0$ s of two waves approaching each other at 1.0 m/s. Draw four snapshot graphs, stacked vertically, showing the string at $t = 2, 4, 6,$ and 8 s.

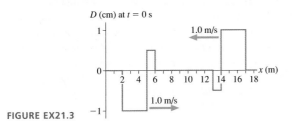

FIGURE EX21.3

4. ‖ **FIGURE EX21.4a** is a snapshot graph at $t = 0$ s of two waves approaching each other at 1.0 m/s.
   a. At what time was the snapshot graph in **FIGURE EX21.4b** taken?
   b. Draw a history graph of the string at $x = 5.0$ m from $t = 0$ s to $t = 6$ s.

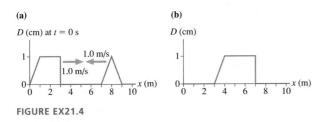

FIGURE EX21.4

Section 21.2 Standing Waves

Section 21.3 Standing Waves on a String

5. ‖ **FIGURE EX21.5** is a snapshot graph at $t = 0$ s of two waves moving to the right at 1.0 m/s. The string is fixed at $x = 8.0$ m. Draw four snapshot graphs, stacked vertically, showing the string at $t = 2, 4, 6,$ and 8 s.

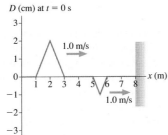

FIGURE EX21.5

6. ‖ **FIGURE EX21.6** shows a standing wave oscillating at 100 Hz on a string. What is the wave speed?

FIGURE EX21.6   **FIGURE EX21.7**

7. ‖ **FIGURE EX21.7** shows a standing wave on a 2.0-m-long string that has been fixed at both ends and tightened until the wave speed is 40 m/s. What is the frequency?

8. ‖ **FIGURE EX21.8** shows a standing wave that is oscillating at frequency f_0.
   a. How many antinodes will there be if the frequency is doubled to $2f_0$? Explain.

FIGURE EX21.8

   b. If the tension in the string is increased by a factor of four, for what frequency, in terms of f_0, will the string continue to oscillate as a standing wave with four antinodes?

9. ‖ a. What are the three longest wavelengths for standing waves on a 240-cm-long string that is fixed at both ends?
   b. If the frequency of the second-longest wavelength is 50 Hz, what is the frequency of the third-longest wavelength?

10. ‖ Standing waves on a 1.0-m-long string that is fixed at both ends are seen at successive frequencies of 36 Hz and 48 Hz.
   a. What are the fundamental frequency and the wave speed?
   b. Draw the standing-wave pattern when the string oscillates at 48 Hz.

11. ‖ A heavy piece of hanging sculpture is suspended by a 90-cm-long, 5.0 g steel wire. When the wind blows hard, the wire hums at its fundamental frequency of 80 Hz. What is the mass of the sculpture?

12. ‖ A carbon dioxide laser is an infrared laser. A CO_2 laser with a cavity length of 53.00 cm oscillates in the $m = 100,000$ mode. What are the wavelength and frequency of the laser beam?

Section 21.4 Standing Sound Waves and Musical Acoustics

13. ‖ What are the three longest wavelengths for standing sound waves in a 121-cm-long tube that is (a) open at both ends and (b) open at one end, closed at the other?

14. ‖ **FIGURE EX21.14** shows a standing sound wave in an 80-cm-long tube. The tube is filled with an unknown gas. What is the speed of sound in this gas?

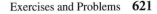

FIGURE EX21.14

15. ‖ The fundamental frequency of an open-open tube is 1500 Hz when the tube is filled with 0°C helium. What is its frequency when filled with 0°C air?

16. ‖ We can make a simple model of the human vocal tract as an open-closed tube extending from the opening of the mouth to the diaphragm. What is the length of this tube if its fundamental frequency equals a typical speech frequency of 250 Hz? The speed of sound in the warm air is 350 m/s.

17. ‖ The lowest note on a grand piano has a frequency of 27.5 Hz. The entire string is 2.00 m long and has a mass of 400 g. The vibrating section of the string is 1.90 m long. What tension is needed to tune this string properly?

18. ‖ A violin string is 30 cm long. It sounds the musical note A (440 Hz) when played without fingering. How far from the end of the string should you place your finger to play the note C (523 Hz)?

Section 21.5 Interference in One Dimension

Section 21.6 The Mathematics of Interference

19. ‖ Two loudspeakers emit sound waves along the x-axis. The sound has maximum intensity when the speakers are 20 cm apart. The sound intensity decreases as the distance between the speakers is increased, reaching zero at a separation of 60 cm.
   a. What is the wavelength of the sound?
   b. If the distance between the speakers continues to increase, at what separation will the sound intensity again be a maximum?

20. ‖ Two loudspeakers in a 20°C room emit 686 Hz sound waves along the x-axis.
    a. If the speakers are in phase, what is the smallest distance between the speakers for which the interference of the sound waves is perfectly destructive?
    b. If the speakers are out of phase, what is the smallest distance between the speakers for which the interference of the sound waves is maximum constructive?

21. | What is the thinnest film of MgF_2 ($n = 1.39$) on glass that produces a strong reflection for orange light with a wavelength of 600 nm?

22. ‖ A very thin oil film ($n = 1.25$) floats on water ($n = 1.33$). What is the thinnest film that produces a strong reflection for green light with a wavelength of 500 nm?

Section 21.7 Interference in Two and Three Dimensions

23. ‖ FIGURE EX21.23 shows the circular wave fronts emitted by two wave sources.
    a. Are these sources in phase or out of phase? Explain.
    b. Make a table with rows labeled P, Q, and R and columns labeled r_1, r_2, Δr, and C/D. Fill in the table for points P, Q, and R, giving the distances as multiples of λ and indicating, with a C or a D, whether the interference at that point is constructive or destructive.

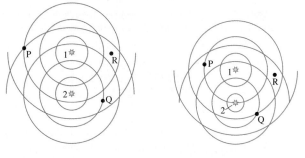

FIGURE EX21.23   FIGURE EX21.24

24. ‖ FIGURE EX21.24 shows the circular wave fronts emitted by two wave sources.
    a. Are these sources in phase or out of phase? Explain.
    b. Make a table with rows labeled P, Q, and R and columns labeled r_1, r_2, Δr, and C/D. Fill in the table for points P, Q, and R, giving the distances as multiples of λ and indicating, with a C or a D, whether the interference at that point is constructive or destructive.

25. ‖ Two in-phase speakers 2.0 m apart in a plane are emitting 1800 Hz sound waves into a room where the speed of sound is 340 m/s. Is the point 4.0 m in front of one of the speakers, perpendicular to the plane of the speakers, a point of maximum constructive interference, perfect destructive interference, or something in between?

26. ‖ Two out-of-phase radio antennas at $x = \pm 300$ m on the x-axis are emitting 3.0 MHz radio waves. Is the point $(x, y) = (300$ m, 800 m$)$ a point of maximum constructive interference, perfect destructive interference, or something in between?

Section 21.8 Beats

27. | Two strings are adjusted to vibrate at exactly 200 Hz. Then the tension in one string is increased slightly. Afterward, three beats per second are heard when the strings vibrate at the same time. What is the new frequency of the string that was tightened?

28. | A flute player hears four beats per second when she compares her note to a 523 Hz tuning fork (the note C). She can match the frequency of the tuning fork by pulling out the "tuning joint" to lengthen her flute slightly. What was her initial frequency?

29. | Two microwave signals of nearly equal wavelengths can generate a beat frequency if both are directed onto the same microwave detector. In an experiment, the beat frequency is 100 MHz. One microwave generator is set to emit microwaves with a wavelength of 1.250 cm. If the second generator emits the longer wavelength, what is that wavelength?

Problems

30. ‖ Two waves on a string travel in opposite directions at 100 m/s. FIGURE P21.30 shows a snapshot graph of the string at $t = 0$ s, when the two waves are overlapped, and a snapshot graph of the left-traveling wave at $t = 0.050$ s. Draw a snapshot graph of the right-traveling wave at $t = 0.050$ s.

FIGURE P21.30

31. | A 2.0-m-long string vibrates at its second-harmonic frequency with a maximum amplitude of 2.0 cm. One end of the string is at $x = 0$ cm. Find the oscillation amplitude at $x = 10, 20, 30, 40,$ and 50 cm.

32. ‖ A string vibrates at its third-harmonic frequency. The amplitude at a point 30 cm from one end is half the maximum amplitude. How long is the string?

33. ‖ A string of length L vibrates at its fundamental frequency. The amplitude at a point $\frac{1}{4}L$ from one end is 2.0 cm. What is the amplitude of each of the traveling waves that form this standing wave?

34. ‖ Two sinusoidal waves with equal wavelengths travel along a string in opposite directions at 3.0 m/s. The time between two successive instants when the antinodes are at maximum height is 0.25 s. What is the wavelength?

35. ‖ BIO Tendons are, essentially, elastic cords stretched between two fixed ends. As such, they can support standing waves. A woman has a 20-cm-long Achilles tendon—connecting the heel to a muscle in the calf—with a cross-section area of 90 mm². The density of tendon tissue is 1100 kg/m³. For a reasonable tension of 500 N, what will be the fundamental frequency of her Achilles tendon?

36. ‖ BIO Biologists think that some spiders "tune" strands of their web to give enhanced response at frequencies corresponding to those at which desirable prey might struggle. Orb spider web silk has a typical diameter of 20 μm, and spider silk has a density of 1300 kg/m³. To have a fundamental frequency at 100 Hz, to what tension must a spider adjust a 12-cm-long strand of silk?

37. ‖ A particularly beautiful note reaching your ear from a rare Stradivarius violin has a wavelength of 39.1 cm. The room is slightly warm, so the speed of sound is 344 m/s. If the string's linear density is 0.600 g/m and the tension is 150 N, how long is the vibrating section of the violin string?

38. ‖ A violinist places her finger so that the vibrating section of a 1.0 g/m string has a length of 30 cm, then she draws her bow across it. A listener nearby in a 20°C room hears a note with a wavelength of 40 cm. What is the tension in the string?

39. ‖ A steel wire is used to stretch the spring of FIGURE P21.39. An oscillating magnetic field drives the steel wire back and forth. A standing wave with three antinodes is created when the spring is stretched 8.0 cm. What stretch of the spring produces a standing wave with two antinodes?

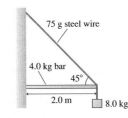

FIGURE P21.39

40. ‖ Astronauts visiting Planet X have a 250-cm-long string whose mass is 5.00 g. They tie the string to a support, stretch it horizontally over a pulley 2.00 m away, and hang a 4.00 kg mass on the free end. Then the astronauts begin to excite standing waves on the horizontal portion of the string. Their data are as follows:

m	Frequency (Hz)
1	31
2	66
3	95
4	130
5	162

Use the best-fit line of an appropriate graph to determine the value of g, the free-fall acceleration on Planet X.

41. ‖ A 75 g bungee cord has an equilibrium length of 1.20 m. The cord is stretched to a length of 1.80 m, then vibrated at 20 Hz. This produces a standing wave with two antinodes. What is the spring constant of the bungee cord?

42. ‖ A metal wire under tension T_0 vibrates at its fundamental frequency. For what tension will the second-harmonic frequency be the same as the fundamental frequency at tension T_0?

43. ‖‖ In a laboratory experiment, one end of a horizontal string is tied to a support while the other end passes over a frictionless pulley and is tied to a 1.5 kg sphere. Students determine the frequencies of standing waves on the horizontal segment of the string, then they raise a beaker of water until the hanging 1.5 kg sphere is completely submerged. The frequency of the fifth harmonic with the sphere submerged exactly matches the frequency of the third harmonic before the sphere was submerged. What is the diameter of the sphere?

44. ‖‖ What is the fundamental frequency of the steel wire in FIGURE P21.44?

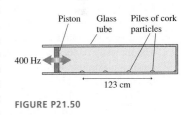

FIGURE P21.44

45. ‖ The two strings in FIGURE P21.45 are of equal length and are being driven at equal frequencies. The linear density of the left string is μ_0. What is the linear density of the right string?

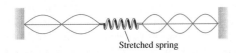

Stretched spring

FIGURE P21.45

46. | Microwaves pass through a small hole into the "microwave cavity" of FIGURE P21.46. What frequencies between 10 GHz and 20 GHz will create standing waves in the cavity?

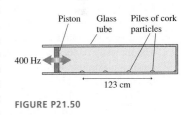

Microwaves

FIGURE P21.46

47. ‖ An open-open organ pipe is 78.0 cm long. An open-closed pipe has a fundamental frequency equal to the third harmonic of the open-open pipe. How long is the open-closed pipe?

48. | A narrow column of 20°C air is found to have standing waves at frequencies of 390 Hz, 520 Hz, and 650 Hz and at no frequencies in between these. The behavior of the tube at frequencies less than 390 Hz or greater than 650 Hz is not known.
   a. Is this an open-open tube or an open-closed tube? Explain.
   b. How long is the tube?

49. ‖ Deep-sea divers often breathe a mixture of helium and oxygen
BIO to avoid getting the "bends" from breathing high-pressure nitrogen. The helium has the side effect of making the divers' voices sound odd. Although your vocal tract can be roughly described as an open-closed tube, the way you hold your mouth and position your lips greatly affects the standing-wave frequencies of the vocal tract. This is what allows different vowels to sound different. The "ee" sound is made by shaping your vocal tract to have standing-wave frequencies at, normally, 270 Hz and 2300 Hz. What will these frequencies be for a helium-oxygen mixture in which the speed of sound at body temperature is 750 m/s? The speed of sound in air at body temperature is 350 m/s.

50. ‖ In 1866, the German scientist Adolph Kundt developed a technique for accurately measuring the speed of sound in various gases. A long glass tube, known today as a Kundt's tube, has a vibrating piston at one end and is closed at the other. Very finely ground particles of cork are sprinkled in the bottom of the tube before the piston is inserted. As the vibrating piston is slowly moved forward, there are a few positions that cause the cork particles to collect in small, regularly spaced piles along the bottom. FIGURE P21.50 shows an experiment in which the tube is filled with pure oxygen and the piston is driven at 400 Hz. What is the speed of sound in oxygen?

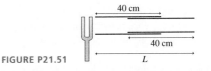

FIGURE P21.50

51. ‖ The 40-cm-long tube of FIGURE P21.51 has a 40-cm-long insert that can be pulled in and out. A vibrating tuning fork is held next to the tube. As the insert is slowly pulled out, the sound from the tuning fork creates standing waves in the tube when the total length L is 42.5 cm, 56.7 cm, and 70.9 cm. What is the frequency of the tuning fork? Assume $v_{sound} = 343$ m/s.

FIGURE P21.51

52. ‖ A 1.0-m-tall vertical tube is filled with 20°C water. A tuning fork vibrating at 580 Hz is held just over the top of the tube as the water is slowly drained from the bottom. At what water heights, measured from the bottom of the tube, will there be a standing wave in the tube above the water?

53. ‖ A 25-cm-long wire with a linear density of 20 g/m passes across the open end of an 85-cm-long open-closed tube of air. If the wire, which is fixed at both ends, vibrates at its fundamental frequency, the sound wave it generates excites the second vibrational mode of the tube of air. What is the tension in the wire? Assume $v_{sound} = 340$ m/s.

54. ‖ A longitudinal standing wave can be created in a long, thin aluminum rod by stroking the rod with very dry fingers. This is often done as a physics demonstration, creating a high-pitched, very annoying whine. From a wave perspective, the standing wave is equivalent to a sound standing wave in an open-open tube. As FIGURE P21.54 shows, both ends of the rod are anti-nodes. What is the fundamental frequency of a 2.0-m-long aluminum rod?

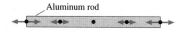

Aluminum rod

FIGURE P21.54

55. ‖ An old mining tunnel disappears into a hillside. You would like to know how long the tunnel is, but it's too dangerous to go inside. Recalling your recent physics class, you decide to try setting up standing-wave resonances inside the tunnel. Using your subsonic amplifier and loudspeaker, you find resonances at 4.5 Hz and 6.3 Hz, and at no frequencies between these. It's rather chilly inside the tunnel, so you estimate the sound speed to be 335 m/s. Based on your measurements, how far is it to the end of the tunnel?

56. ‖ Analyze the standing sound waves in an open-closed tube to show that the possible wavelengths and frequencies are given by Equation 21.18.

57. ‖‖ Two in-phase loudspeakers emit identical 1000 Hz sound waves along the *x*-axis. What distance should one speaker be placed behind the other for the sound to have an amplitude 1.5 times that of each speaker alone?

58. ‖ Two loudspeakers emit sound waves of the same frequency along the *x*-axis. The amplitude of each wave is *a*. The sound intensity is minimum when speaker 2 is 10 cm behind speaker 1. The intensity increases as speaker 2 is moved forward and first reaches maximum, with amplitude 2*a*, when it is 30 cm in front of speaker 1. What is
   a. The wavelength of the sound?
   b. The phase difference between the two loudspeakers?
   c. The amplitude of the sound (as a multiple of *a*) if the speakers are placed side by side?

59. ‖‖ Two loudspeakers emit sound waves along the *x*-axis. A listener in front of both speakers hears a maximum sound intensity when speaker 2 is at the origin and speaker 1 is at *x* = 0.50 m. If speaker 1 is slowly moved forward, the sound intensity decreases and then increases, reaching another maximum when speaker 1 is at *x* = 0.90 m.
   a. What is the frequency of the sound? Assume $v_{sound} = 340$ m/s.
   b. What is the phase difference between the speakers?

60. ‖ A sheet of glass is coated with a 500-nm-thick layer of oil (*n* = 1.42).
   a. For what *visible* wavelengths of light do the reflected waves interfere constructively?
   b. For what *visible* wavelengths of light do the reflected waves interfere destructively?
   c. What is the color of reflected light? What is the color of transmitted light?

61. ‖ A manufacturing firm has hired your company, Acoustical Consulting, to help with a problem. Their employees are complaining about the annoying hum from a piece of machinery. Using a frequency meter, you quickly determine that the machine emits a rather loud sound at 1200 Hz. After investigating, you tell the owner that you cannot solve the problem entirely, but you can at least improve the situation by eliminating reflections of this sound from the walls. You propose to do this by installing mesh screens in front of the walls. A portion of the sound will reflect from the mesh; the rest will pass through the mesh and reflect from the wall. How far should the mesh be placed in front of the wall for this scheme to work?

62. ‖ A soap bubble is essentially a very thin film of water (*n* = 1.33) surrounded by air. The colors that you see in soap bubbles are produced by interference.
   a. Derive an expression for the wavelengths λ_C for which constructive interference causes a strong reflection from a soap bubble of thickness *d*.
   Hint: Think about the reflection phase shifts at both boundaries.
   b. What visible wavelengths of light are strongly reflected from a 390-nm-thick soap bubble? What color would such a soap bubble appear to be?

63. ‖ Two radio antennas are separated by 2.0 m. Both broadcast identical 750 MHz waves. If you walk around the antennas in a circle of radius 10 m, how many maxima will you detect?

64. ‖ You are standing 2.5 m directly in front of one of the two loudspeakers shown in FIGURE P21.64. They are 3.0 m apart and both are playing a 686 Hz tone in phase. As you begin to walk directly away from the speaker, at what distances from the speaker do you hear a *minimum* sound intensity? The room temperature is 20°C.

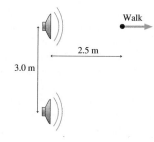

Walk

2.5 m

3.0 m

FIGURE P21.64

65. ‖ Two loudspeakers in a plane, 5.0 apart, are playing the same frequency. If you stand 12.0 m in front of the plane of the speakers, centered between them, you hear a sound of maximum intensity. As you walk parallel to the plane of the speakers, staying 12.0 m in front of them, you first hear a minimum of sound intensity when you are directly in front of one of the speakers. What is the frequency of the sound? Assume a sound speed of 340 m/s.

66. ‖ Two in-phase loudspeakers are located at (*x*, *y*) coordinates (−3.0 m, +2.0 m) and (−3.0 m, −2.0 m). They emit identical sound waves with a 2.0 m wavelength and amplitude *a*. Determine the amplitude of the sound at the five positions on the *y*-axis (*x* = 0) with *y* = 0.0 m, 0.5 m, 1.0 m, 1.5 m, and 2.0 m.

67. ‖ Two identical loudspeakers separated by distance Δ*x* each emit sound waves of wavelength λ and amplitude *a* along the *x*-axis. What is the minimum value of the ratio Δ*x*/λ for which the amplitude of their superposition is also *a*?

68. ‖ Two radio antennas are 100 m apart along a north-south line. They broadcast identical radio waves at a frequency of 3.0 MHz. Your job is to monitor the signal strength with a hand-held receiver. To get to your first measuring point, you walk 800 m east from the midpoint between the antennas, then 600 m north.
    a. What is the phase difference between the waves at this point?
    b. Is the interference at this point maximum constructive, perfect destructive, or somewhere in between? Explain.
    c. If you now begin to walk farther north, does the signal strength increase, decrease, or stay the same? Explain.

69. ‖ The three identical loudspeakers in **FIGURE P21.69** play a 170 Hz tone in a room where the speed of sound is 340 m/s. You are standing 4.0 m in front of the middle speaker. At this point, the amplitude of the wave from each speaker is a.
    a. What is the amplitude at this point?
    b. How far must speaker 2 be moved to the left to produce a maximum amplitude at the point where you are standing?
    c. When the amplitude is maximum, by what factor is the sound intensity greater than the sound intensity from a single speaker?

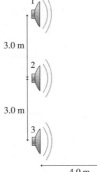

FIGURE P21.69

3.0 m

3.0 m

4.0 m

70. | Piano tuners tune pianos by listening to the beats between the *harmonics* of two different strings. When properly tuned, the note A should have a frequency of 440 Hz and the note E should be at 659 Hz.
    a. What is the frequency difference between the third harmonic of the A and the second harmonic of the E?
    b. A tuner first tunes the A string very precisely by matching it to a 440 Hz tuning fork. She then strikes the A and E strings simultaneously and listens for beats between the harmonics. What beat frequency indicates that the E string is properly tuned?
    c. The tuner starts with the tension in the E string a little low, then tightens it. What is the frequency of the E string when she hears four beats per second?

71. ‖ A flutist assembles her flute in a room where the speed of sound is 342 m/s. When she plays the note A, it is in perfect tune with a 440 Hz tuning fork. After a few minutes, the air inside her flute has warmed to where the speed of sound is 346 m/s.
    a. How many beats per second will she hear if she now plays the note A as the tuning fork is sounded?
    b. How far does she need to extend the "tuning joint" of her flute to be in tune with the tuning fork?

72. ‖ Two loudspeakers face each other from opposite walls of a room. Both are playing exactly the same frequency, thus setting up a standing wave with distance $\lambda/2$ between antinodes. Assume that λ is much less than the room width, so there are many antinodes.
    a. Yvette starts at one speaker and runs toward the other at speed v_Y. As the does so, she hears a loud-soft-loud modulation of the sound intensity. From your perspective, as you sit at rest in the room, Yvette is running through the nodes and antinodes of the standing wave. Find an expression for the number of sound maxima she hears per second.

    b. From Yvette's perspective, the two sound waves are Doppler shifted. They're not the same frequency, so they don't create a standing wave. Instead, she hears a loud-soft-loud modulation of the sound intensity because of beats. Find an expression for the beat frequency that Yvette hears.
    c. Are your answers to parts a and b the same or different? *Should* they be the same or different?

73. ‖ Two loudspeakers emit 400 Hz notes. One speaker sits on the ground. The other speaker is in the back of a pickup truck. You hear eight beats per second as the truck drives away from you. What is the truck's speed?

Challenge Problems

74. a. The frequency of a standing wave on a string is f when the string's tension is T_s. If the tension is changed by the *small* amount ΔT_s, without changing the length, show that the frequency changes by an amount Δf such that

$$\frac{\Delta f}{f} = \frac{1}{2}\frac{\Delta T_s}{T_s}$$

   b. Two identical strings vibrate at 500 Hz when stretched with the same tension. What percentage increase in the tension of one of the strings will cause five beats per second when both strings vibrate simultaneously?

75. A 280 Hz sound wave is directed into one end of the trombone slide seen in **FIGURE CP21.75**. A microphone is placed at the other end to record the intensity of sound waves that are transmitted through the tube. The straight sides of the slide are 80 cm in length and 10 cm apart with a semicircular bend at the end. For what slide extensions s will the microphone detect a maximum of sound intensity?

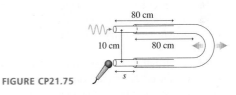

80 cm

10 cm

80 cm

s

FIGURE CP21.75

76. As the captain of the scientific team sent to Planet Physics, one of your tasks is to measure g. You have a long, thin wire labeled 1.00 g/m and a 1.25 kg weight. You have your accurate space cadet chronometer but, unfortunately, you seem to have forgotten a meter stick. Undeterred, you first find the midpoint of the wire by folding it in half. You then attach one end of the wire to the wall of your laboratory, stretch it horizontally to pass over a pulley at the midpoint of the wire, then tie the 1.25 kg weight to the end hanging over the pulley. By vibrating the wire, and measuring time with your chronometer, you find that the wire's second-harmonic frequency is 100 Hz. Next, with the 1.25 kg weight still tied to one end of the wire, you attach the other end to the ceiling to make a pendulum. You find that the pendulum requires 314 s to complete 100 oscillations. Pulling out your trusty calculator, you get to work. What value of g will you report back to headquarters?

77. When mass M is tied to the bottom of a long, thin wire suspended from the ceiling, the wire's second-harmonic frequency is 200 Hz. Adding an additional 1.0 kg to the hanging mass increases the second-harmonic frequency to 245 Hz. What is M?

78. Ultrasound has many medical applications, one of which is to
BIO monitor fetal heartbeats by reflecting ultrasound off a fetus in the
womb.
   a. Consider an object moving at speed v_o toward an at-rest
   source that is emitting sound waves of frequency f_0. Show
   that the reflected wave (i.e., the echo) that returns to the
   source has a Doppler-shifted frequency

$$f_{echo} = \left(\frac{v + v_o}{v - v_o}\right) f_0$$

   where v is the speed of sound in the medium.
   b. Suppose the object's speed is much less than the wave speed:
   $v_o \ll v$. Then $f_{echo} \approx f_0$, and a microphone that is sensitive to
   these frequencies will detect a beat frequency if it listens to
   f_0 and f_{echo} simultaneously. Use the binomial approximation
   and other appropriate approximations to show that the beat
   frequency is $f_{beat} \approx (2v_o/v)f_0$.
   c. The reflection of 2.40 MHz ultrasound waves from the surface
   of a fetus's beating heart is combined with the 2.40 MHz wave
   to produce a beat frequency that reaches a maximum of 65 Hz.
   What is the maximum speed of the surface of the heart? The
   speed of ultrasound waves within the body is 1540 m/s.
   d. Suppose the surface of the heart moves in simple harmonic
   motion at 90 beats/min. What is the amplitude in mm of the
   heartbeat?

79. A water wave is called a *deep-water wave* if the water's depth
is more than one-quarter of the wavelength. Unlike the waves
we've considered in this chapter, the speed of a deep-water wave
depends on its wavelength:

$$v = \sqrt{\frac{g\lambda}{2\pi}}$$

Longer wavelengths travel faster. Let's apply this to standing
waves. Consider a diving pool that is 5.0 m deep and 10.0 m
wide. Standing water waves can set up across the width of the
pool. Because water sloshes up and down at the sides of the pool,
the boundary conditions require antinodes at $x = 0$ and $x = L$.
Thus a standing water wave resembles a standing sound wave in
an open-open tube.
   a. What are the wavelengths of the first three standing-wave
   modes for water in the pool? Do they satisfy the condition for
   being deep-water waves? Draw a graph of each.

   b. What are the wave speeds for each of these waves?
   c. Derive a general expression for the frequencies f_m of the possible standing waves. Your expression should be in terms of
   m, g, and L.
   d. What are the oscillation *periods* of the first three standing-
   wave modes?

80. The broadcast antenna of an
AM radio station is located at
the edge of town. The station
owners would like to beam all
of the energy into town and
none into the countryside, but
a single antenna radiates energy equally in all directions.
FIGURE CP21.80 shows two parallel antennas separated by dis-
tance L. Both antennas broadcast a signal at wavelength λ, but
antenna 2 can delay its broadcast relative to antenna 1 by a time
interval Δt in order to create a phase difference $\Delta\phi_0$ between
the sources. Your task is to find values of L and Δt such that the
waves interfere constructively on the town side and destructively
on the country side.

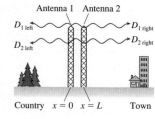

FIGURE CP21.80

Let antenna 1 be at $x = 0$. The wave that travels to the right is
$a\sin[2\pi(x/\lambda - t/T)]$. The left wave is $a\sin[2\pi(-x/\lambda - t/T)]$.
(It must be this, rather than $a\sin[2\pi(x/\lambda + t/T)]$, so that the
two waves match at $x = 0$.) Antenna 2 is at $x = L$. It broadcasts
wave $a\sin[2\pi((x - L)/\lambda - t/T) + \phi_{20}]$ to the right and wave
$a\sin[2\pi(-(x - L)/\lambda - t/T) + \phi_{20}]$ to the left.
   a. What is the smallest value of L for which you can create perfect constructive interference on the town side and perfect
   destructive interference on the country side? Your answer
   will be a multiple or fraction of the wavelength λ.
   b. What phase constant ϕ_{20} of antenna 2 is needed?
   c. What fraction of the oscillation period T must Δt be to produce the proper value of ϕ_{20}?
   d. Evaluate both L and Δt for the realistic AM radio frequency
   of 1000 KHz.

Comment: This is a simple example of what is called a *phased
array*, where phase differences between identical emitters are
used to "steer" the radiation in a particular direction. Phased ar-
rays are widely used in radar technology.

STOP TO THINK ANSWERS

Stop to Think 21.1: c. The figure shows the two waves at $t = 6$ s and
their superposition. The superposition is the *point-by-point* addition
of the displacements of the two individual waves.

$$x \text{ (m)}$$
0 2 4 6 8 10 12 14 16 18 20

Stop to Think 21.2: a. The allowed standing-wave frequencies are
$f_m = m(v/2L)$, so the mode number of a standing wave of frequency
f is $m = 2Lf/v$. Quadrupling T_s increases the wave speed v by a factor
of 2. The initial mode number was 2, so the new mode number is 1.

Stop to Think 21.3: b. 300 Hz and 400 Hz are allowed standing
waves, but they are not f_1 and f_2 because 400 Hz $\neq 2 \times 300$ Hz.
Because there's a 100 Hz difference between them, these must be

$f_3 = 3 \times 100$ Hz and $f_4 = 4 \times 100$ Hz, with a fundamental frequency
$f_1 = 100$ Hz. Thus the second harmonic is $f_2 = 2 \times 100$ Hz $=$
200 Hz.

Stop to Think 21.4: c. Shifting the top wave 0.5 m to the left aligns
crest with crest and trough with trough.

Stop to Think 21.5: a. $r_1 = 0.5\lambda$ and $r_2 = 2.5\lambda$, so $\Delta r = 2.0\lambda$. This
is the condition for maximum constructive interference.

Stop to Think 21.6: Maximum constructive. The path-length difference is $\Delta r = 1.0$ m $= \lambda$. For identical sources, interference is constructive when Δr is an integer multiple of λ.

Stop to Think 21.7: f. The beat frequency is the difference between
the two frequencies.

22 Wave Optics

The vivid colors of this peacock—which change as you see the feathers from different angles—are not due to pigments. Instead, the colors are due to the interference of light waves.

▶ **Looking Ahead** The goal of Chapter 22 is to understand and apply the wave model of light.

Models of Light

You'll learn that light has aspects of both waves and particles. We'll introduce three models of light:

The **wave model** of light—the subject of this chapter—allows us to understand the colors of a soap bubble.

To understand the focusing of light by a contact lens, Chapter 23 will introduce a **ray model** in which light travels in particle-like straight lines.

Solar cells generate electricity from sunlight. The **photon model** of Part VII will be most appropriate for understanding this aspect of light.

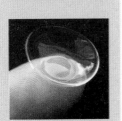

◀ **Looking Back**
Sections 20.4–20.6 Wave fronts, phase, and intensity

Diffraction

Diffraction is the ability of waves to spread out after going through small holes or around corners. The diffraction of light indicates that light is a wave.

The "ripples" around the edges of this razor blade—back lit with a blue laser beam—are due to the diffraction of light.

The Diffraction Grating

A **diffraction grating** is a periodic array of closely spaced holes or slits or grooves. You'll learn how a diffraction grating sends different wavelengths off at different angles.

The microscopic pits in this DVD act as a diffraction grating, breaking white light into its component colors.

Diffraction gratings are the basis for *spectroscopy*, an important tool for determining the composition of materials by the wavelengths they emit.

Double-Slit Interference

You'll learn that an interference pattern is formed when light shines on an opaque screen with two narrow, closely spaced slits. This also shows that light is a wave.

Interference fringes from green light passing through two closely spaced slits

◀ **Looking Back**
Section 21.7 Interference

Interferometry

Today, the controlled interference of light has applications that include optical computing, precision measurements in engineering, holography, and observing movements of the earth's crust.

Interference fringes such as these can be used to monitor vibrations and displacements of only a few nanometers.

22.1 Light and Optics

The study of light is called **optics.** But what is light? The first Greek scientists did not make a distinction between light and vision. Light, to them, was inseparable from seeing. But gradually there arose a view that light actually "exists," that light is some sort of physical entity that is present regardless of whether or not someone is looking. But if light is a physical entity, what is it? What are its characteristics? Is it a wave, similar to sound? Or is light a collection of small particles that blows by like the wind?

Newton, in addition to his pioneering work in mathematics and mechanics in the 1660s, investigated the nature of light. Newton knew that a water wave, after passing through an opening, *spreads out* to fill the space behind the opening. You can see this in FIGURE 22.1a, where plane waves, approaching from the left, spread out in circular arcs after passing through a hole in a barrier. This inexorable spreading of waves is the phenomenon called **diffraction.** Diffraction is a sure sign that whatever is passing through the hole is a wave.

In contrast, FIGURE 22.1b shows that sunlight makes a sharp-edged shadow after passing through a door. We don't see sunlight light spreading out in circular arcs. This behavior is exactly what you would expect if light consists of particles traveling in straight lines. Some particles would pass through the door to make a bright area on the floor, others would be blocked and cause the well-defined shadow. This reasoning led Newton to the conclusion that light consists of very small, light, fast particles that he called *corpuscles.*

The situation changed dramatically in 1801, when the English scientist Thomas Young announced that he had produced *interference* between two waves of light. Young's experiment, which we will analyze in the next section, was painstakingly difficult with the technology of his era. Nonetheless, Young's experiment quickly settled the debate in favor of a wave theory of light because interference is a distinctly wave-like phenomenon.

But if light is a wave, what is waving? This was the question that Young posed to the 19th century. It was ultimately established that light is an *electromagnetic wave,* an oscillation of the electromagnetic field requiring no material medium in which to travel. Further, as we have already seen, visible light is just one small slice out of a vastly broader *electromagnetic spectrum.*

But this satisfying conclusion was soon undermined by new discoveries at the start of the 20th century. Albert Einstein's introduction of the concept of the *photon*—a wave having certain particle-like characteristics—marked the end of *classical physics* and the beginning of a new era called *quantum physics.* Equally important, Einstein's theory marked yet another shift in our age-old effort to understand light.

FIGURE 22.1 Water waves spread out behind a small hole in a barrier, but light passing through a doorway makes a sharp-edged shadow.

(a) Plane waves approach from the left.

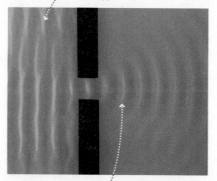

Circular waves spread out on the right.

(b)

A beam of sunlight has a sharp edge.

Models of Light

Light is a real physical entity, but the nature of light is elusive. Light is the chameleon of the physical world. Under some circumstances, light acts like particles traveling in straight lines. But change the circumstances, and light shows the same kinds of wave-like behavior as sound waves or water waves. Change the circumstances yet again, and light exhibits behavior that is neither wave-like nor particle-like but has characteristics of both.

Rather than an all-encompassing "theory of light," it will be better to develop three **models of light.** Each model successfully explains the behavior of light within a certain domain—that is, within a certain range of physical situations. Our task will be twofold:

1. To develop clear and distinct models of light.
2. To learn the conditions and circumstances for which each model is valid.

We'll begin with a brief summary of all three models.

Three models of light

The Wave Model

The wave model of light is responsible for the widely known "fact" that light is a wave. Indeed, under many circumstances light exhibits the same behavior as sound or water waves. Lasers and electro-optical devices are best described by the wave model of light. Some aspects of the wave model were introduced in Chapters 20 and 21, and it is the primary focus of this chapter.

The Ray Model

An equally well-known "fact" is that light travels in straight lines. These straight-line paths are called *light rays*. The properties of prisms, mirrors, and lenses are best understood in terms of light rays. Unfortunately, it's difficult to reconcile "light travels in straight lines" with "light is a wave." For the most part, waves and rays are mutually exclusive models of light. One of our important tasks will be to learn when each model is appropriate. Ray optics is the subject of Chapters 23 and 24.

The Photon Model

Modern technology is increasingly reliant on quantum physics. In the quantum world, light behaves like neither a wave nor a particle. Instead, light consists of *photons* that have both wave-like and particle-like properties. Much of the quantum theory of light is beyond the scope of this textbook, but we will take a peek at some of the important ideas in Part VII.

22.2 The Interference of Light

Newton might have reached a different conclusion had he seen the experiment depicted in FIGURE 22.2. Here light of a single wavelength (or color) passes through a "window"—a narrow slit—that is only 0.1 mm wide, about twice the width of a human hair. The image shows how the light appears on a viewing screen 2 m behind the slit. If light consists of corpuscles traveling in straight lines, as Newton thought, we should see a narrow strip of light, about 0.1 mm wide, with dark shadows on either side. Instead, we see a band of light extending over about 2.5 cm, a distance much wider than the aperture, with dimmer patches of light extending even farther on either side.

If you compare Figure 22.2 to the water wave of Figure 22.1, you see that *the light is spreading out* behind the 0.1-mm-wide hole. The light is exhibiting diffraction, the sure signature of waviness. We will look at diffraction in more detail later in the chapter. For now, we merely need the *observation* that light does, indeed, spread out behind a hole that is sufficiently small.

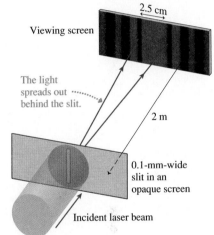

FIGURE 22.2 Light, just like a water wave, does spread out behind a hole *if* the hole is sufficiently small.

Young's Double-Slit Experiment

Rather than one small hole, suppose we use two. FIGURE 22.3a shows an experiment in which a laser beam is aimed at an opaque screen containing two long, narrow slits that are very close together. This pair of slits is called a **double slit,** and in a typical

FIGURE 22.3 A double-slit interference experiment.

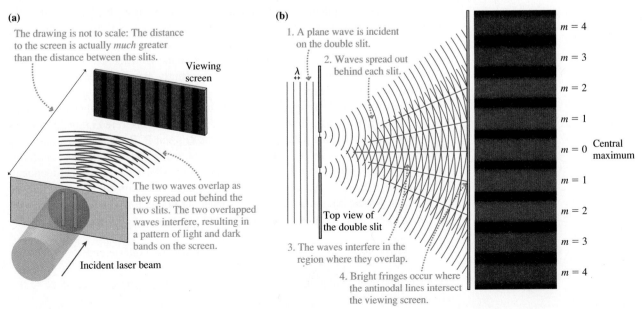

(a) The drawing is not to scale: The distance to the screen is actually *much* greater than the distance between the slits.

Viewing screen

The two waves overlap as they spread out behind the two slits. The two overlapped waves interfere, resulting in a pattern of light and dark bands on the screen.

Incident laser beam

(b) 1. A plane wave is incident on the double slit.

2. Waves spread out behind each slit.

λ

Top view of the double slit

3. The waves interfere in the region where they overlap.

4. Bright fringes occur where the antinodal lines intersect the viewing screen.

$m = 4$
$m = 3$
$m = 2$
$m = 1$
$m = 0$ Central maximum
$m = 1$
$m = 2$
$m = 3$
$m = 4$

experiment they are ≈ 0.1 mm wide and spaced ≈ 0.5 mm apart. We will assume that the laser beam illuminates both slits equally, and any light passing through the slits impinges on a viewing screen. This is the essence of Young's experiment of 1801, although he used sunlight rather than a laser.

What should we expect to see on the screen? FIGURE 22.3b is a view from above the experiment, looking down on the top ends of the slits and the top edge of the viewing screen. Because the slits are very narrow, **light spreads out behind each slit** as it did in Figure 22.2, and these two spreading waves overlap in the region between the slits and the screen.

The primary conclusion of Chapter 21 was that two overlapped waves of equal wavelength produce interference. In fact, Figure 22.3b is equivalent to the waves emitted by two loudspeakers, a situation we analyzed in Section 21.7. (It is very useful to compare Figure 22.3b with Figures 21.30 and 21.32a.) Nothing in that analysis depended on what type of wave it was, so the conclusions apply equally well to two overlapped light waves. If light really is a wave, we should see interference between the two light waves over the small region, typically a few centimeters wide, where they overlap on the viewing screen.

The image in Figure 22.3b shows how the screen looks. As expected, the light is intense at points where an antinodal line intersects the screen. There is no light at all at points where a nodal line intersects the screen. These alternating bright and dark bands of light, due to constructive and destructive interference, are called **interference fringes**. The fringes are numbered $m = 0, 1, 2, 3, \ldots$, going outward from the center. The brightest fringe, at the midpoint of the viewing screen, with $m = 0$, is called the **central maximum.**

STOP TO THINK 22.1 Suppose the viewing screen in Figure 22.3 is moved closer to the double slit. What happens to the interference fringes?

a. They get brighter but otherwise do not change.
b. They get brighter and closer together.
c. They get brighter and farther apart.
d. They get out of focus.
e. They fade out and disappear.

Analyzing Double-Slit Interference

Figure 22.3 showed qualitatively how interference is produced behind a double slit by the overlap of the light waves spreading out behind each slit. Now let's analyze the experiment more carefully. FIGURE 22.4 shows a double-slit experiment in which the spacing between the two slits is d and the distance to the viewing screen is L. We will assume that L is *very* much larger than d. Consequently, we don't see the individual slits in the upper part of Figure 22.4.

Let P be a point on the screen at angle θ. Our goal is to determine whether the interference at P is constructive, destructive, or in between. The insert to Figure 22.4 shows the individual slits and the paths from these slits to point P. Because P is so far away on this scale, the two paths are virtually parallel, both at angle θ. Both slits are illuminated by the *same* wave front from the laser; hence the slits act as sources of identical, in-phase waves ($\Delta\phi_0 = 0$). You learned in Chapter 21 that constructive interference between the waves from in-phase sources occurs at points for which the path-length difference $\Delta r = r_2 - r_1$ is an integer number of wavelengths:

$$\Delta r = m\lambda \qquad m = 0, 1, 2, 3, \ldots \qquad \text{(constructive interference)} \quad (22.1)$$

Thus the interference at point P is constructive, producing a bright fringe, if $\Delta r = m\lambda$ at that point.

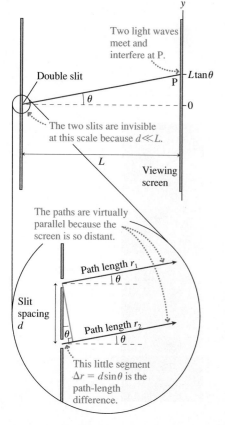

FIGURE 22.4 Geometry of the double-slit experiment.

Double slit

Two light waves meet and interfere at P.

$L\tan\theta$

P

θ

0

The two slits are invisible at this scale because $d \ll L$.

L

Viewing screen

The paths are virtually parallel because the screen is so distant.

Path length r_1

θ

Slit spacing d

θ

Path length r_2

θ

This little segment $\Delta r = d\sin\theta$ is the path-length difference.

The midpoint on the viewing screen at $y = 0$ is equally distant from both slits ($\Delta r = 0$) and thus is a point of constructive interference. This is the bright fringe identified as the central maximum in Figure 22.3b. The path-length difference increases as you move away from the center of the screen, and the $m = 1$ fringes occur at the points where $\Delta r = 1\lambda$—that is, where one wave has traveled exactly one wavelength farther than the other. In general, **the mth bright fringe occurs where the wave from one slit travels m wavelengths farther than the wave from the other slit and thus $\Delta r = m\lambda$.**

You can see from the magnified portion of Figure 22.4 that the wave from the lower slit travels an extra distance

$$\Delta r = d \sin\theta \qquad (22.2)$$

If we use this in Equation 22.1, we find that bright fringes (constructive interference) occur at angles θ_m such that

$$\Delta r = d \sin\theta_m = m\lambda \qquad m = 0, 1, 2, 3, \ldots \qquad (22.3)$$

We added the subscript m to denote that θ_m is the angle of the mth bright fringe, starting with $m = 0$ at the center.

In practice, the angle θ in a double-slit experiment is very small ($<1°$). We can use the small-angle approximation $\sin\theta \approx \theta$, where θ must be in radians, to write Equation 22.3 as

$$\theta_m = m\frac{\lambda}{d} \qquad m = 0, 1, 2, 3, \ldots \qquad \text{(angles of bright fringes)} \qquad (22.4)$$

This gives the angular positions *in radians* of the bright fringes in the interference pattern.

It's usually easier to measure distances rather than angles, so we can also specify point P by its position on a y-axis with the origin directly across from the midpoint between the slits. You can see from Figure 22.4 that

$$y = L \tan\theta \qquad (22.5)$$

Using the small-angle approximation once again, this time in the form $\tan\theta \approx \theta$, we can substitute θ_m from Equation 22.4 for $\tan\theta_m$ in Equation 22.1 to find that the mth bright fringe occurs at position

$$y_m = \frac{m\lambda L}{d} \qquad m = 0, 1, 2, 3, \ldots \qquad \text{(positions of bright fringes)} \quad (22.6)$$

The interference pattern is symmetrical, so there is an mth bright fringe at the same distance on both sides of the center. You can see this in Figure 22.3b. As we've noted, **the $m = 1$ fringes occur at points on the screen where the light from one slit travels exactly one wavelength farther than the light from the other slit.**

NOTE ▶ Equations 22.4 and 22.6 do *not* apply to the interference of sound waves from two loudspeakers. The approximations we've used (small angles, $L \gg d$) are usually not valid for the much longer wavelengths of sound waves. ◀

Equation 22.6 predicts that **the interference pattern is a series of equally spaced bright lines** on the screen, exactly as shown in Figure 22.3b. How do we know the fringes are equally spaced? The **fringe spacing** between the m fringe and the $m + 1$ fringe is

$$\Delta y = y_{m+1} - y_m = \frac{(m + 1)\lambda L}{d} - \frac{m\lambda L}{d} = \frac{\lambda L}{d} \qquad (22.7)$$

Because Δy is independent of m, *any* two adjacent bright fringes have the same spacing.

The dark fringes in the image are bands of destructive interference. You learned in Chapter 21 that destructive interference occurs at positions where the path-length difference of the waves is a half-integer number of wavelengths:

$$\Delta r = \left(m + \frac{1}{2}\right)\lambda \qquad m = 0, 1, 2, \ldots \qquad \text{(destructive interference)} \qquad (22.8)$$

We can use Equation 22.2 for Δr and the small-angle approximation to find that the dark fringes are located at positions

$$y'_m = \left(m + \frac{1}{2}\right)\frac{\lambda L}{d} \qquad m = 0, 1, 2, \ldots \qquad \text{(positions of dark fringes)} \qquad (22.9)$$

We have used y'_m, with a prime, to distinguish the location of the mth minimum from the mth maximum at y_m. You can see from Equation 22.9 that **the dark fringes are located exactly halfway between the bright fringes.**

EXAMPLE 22.1 **Double-slit interference of a laser beam**

Light from a helium-neon laser ($\lambda = 633$ nm) illuminates two slits spaced 0.40 mm apart. A viewing screen is 2.0 m behind the slits. What are the distances between the two $m = 2$ bright fringes and between the two $m = 2$ dark fringes?

MODEL Two closely spaced slits produce a double-slit interference pattern.

VISUALIZE The interference pattern looks like the image of Figure 22.3b. It is symmetrical, with $m = 2$ bright fringes at equal distances on both sides of the central maximum.

SOLVE The positions of the bright fringes are given by Equation 22.6. The $m = 2$ bright fringe is located at position

$$y_m = \frac{m\lambda L}{d} = \frac{2(633 \times 10^{-9}\,\text{m})(2.0\,\text{m})}{4.0 \times 10^{-4}\,\text{m}} = 6.3\,\text{mm}$$

Each of the $m = 2$ fringes is 6.3 mm from the central maximum; so the distance between the two $m = 2$ bright fringes is 12.6 mm. The $m = 2$ dark fringe is located at

$$y'_m = \left(m + \frac{1}{2}\right)\frac{\lambda L}{d} = 7.9\,\text{mm}$$

Thus the distance between the two $m = 2$ dark fringes is 15.8 mm.

ASSESS Because the fringes are counted outward from the center, the $m = 2$ bright fringe occurs *before* the $m = 2$ dark fringe.

EXAMPLE 22.2 **Measuring the wavelength of light**

A double-slit interference pattern is observed on a screen 1.0 m behind two slits spaced 0.30 mm apart. Ten bright fringes span a distance of 1.7 cm. What is the wavelength of the light?

MODEL It is not always obvious which fringe is the central maximum. Slight imperfections in the slits can make the interference fringe pattern less than ideal. However, you do not need to identify the $m = 0$ fringe because you can make use of the fact that the fringe spacing Δy is uniform. Ten bright fringes have *nine* spaces between them (not ten—be careful!).

VISUALIZE The interference pattern looks like the image of Figure 22.3b.

SOLVE The fringe spacing is

$$\Delta y = \frac{1.7\,\text{cm}}{9} = 1.89 \times 10^{-3}\,\text{m}$$

Using this fringe spacing in Equation 22.7, we find that the wavelength is

$$\lambda = \frac{d}{L}\Delta y = 5.7 \times 10^{-7}\,\text{m} = 570\,\text{nm}$$

It is customary to express the wavelengths of visible light in nanometers. Be sure to do this as you solve problems.

ASSESS Young's double-slit experiment not only demonstrated that light is a wave, it provided a means for measuring the wavelength. You learned in Chapter 20 that the wavelengths of visible light span the range 400–700 nm. These lengths are smaller than we can easily comprehend. A wavelength of 570 nm, which is in the middle of the visible spectrum, is only about 1% of the diameter of a human hair.

STOP TO THINK 22.2 Light of wavelength λ_1 illuminates a double slit, and interference fringes are observed on a screen behind the slits. When the wavelength is changed to λ_2, the fringes get closer together. Is λ_2 larger or smaller than λ_1?

Intensity of the Double-Slit Interference Pattern

Equations 22.6 and 22.9 locate the positions of maximum and zero intensity. To complete our analysis we need to calculate the light *intensity* at every point on the screen. All the tools we need to do this calculation were developed in Chapters 20 and 21.

You learned in Chapter 20 that the wave intensity I is proportional to the square of the wave's amplitude. The light spreading out behind a *single* slit produces the wide band of light that you saw in Figure 22.2. The intensity in this band of light is $I_1 = ca^2$, where a is the light-wave amplitude at the screen due to *one* wave and c is a proportionality constant.

If there were no interference, the light intensity due to two slits would be twice the intensity of one slit: $I_2 = 2I_1 = 2ca^2$. In other words, two slits would cause the broad band of light on the screen to be twice as bright. But that's not what happens. Instead, the superposition of the two light waves creates bright and dark interference fringes.

We found in Chapter 21 (Equation 21.36) that the net amplitude of two superimposed waves is

$$A = \left| 2a \cos\left(\frac{\Delta\phi}{2}\right) \right| \tag{22.10}$$

where a is the amplitude of each individual wave. Because the sources (i.e., the two slits) are in phase, the phase difference $\Delta\phi$ at the point where the two waves are combined is due only to the path-length difference: $\Delta\phi = 2\pi(\Delta r/\lambda)$. Using Equation 22.2 for Δr, along with the small-angle approximation and Equation 22.5 for y, we find the phase difference at position y on the screen to be

$$\Delta\phi = 2\pi\frac{\Delta r}{\lambda} = 2\pi\frac{d\sin\theta}{\lambda} \approx 2\pi\frac{d\tan\theta}{\lambda} = \frac{2\pi d}{\lambda L}y \tag{22.11}$$

Substituting Equation 22.11 into Equation 22.10, we find the wave amplitude at position y to be

$$A = \left| 2a \cos\left(\frac{\pi d}{\lambda L}y\right) \right| \tag{22.12}$$

Consequently, the light intensity at position y on the screen is

$$I = cA^2 = 4ca^2 \cos^2\left(\frac{\pi d}{\lambda L}y\right) \tag{22.13}$$

But ca^2 is I_1, the light intensity of a single slit. Thus the intensity of the double-slit interference pattern at position y is

$$I_{\text{double}} = 4I_1 \cos^2\left(\frac{\pi d}{\lambda L}y\right) \tag{22.14}$$

FIGURE 22.5a is a graph of the double-slit intensity versus position y. Notice the unusual orientation of the graph, with the intensity increasing toward the *left* so that the y-axis can match the experimental layout. You can see that the intensity oscillates between dark fringes ($I_{\text{double}} = 0$) and bright fringes ($I_{\text{double}} = 4I_1$). The maxima occur at points where $y_m = m\lambda L/d$. This is what we found earlier for the positions of the bright fringes, so Equation 22.14 is consistent with our initial analysis.

One curious feature is that the light intensity at the maxima is $I = 4I_1$, four times the intensity of the light from each slit alone. You might think that two slits would make the light twice as intense as one slit, but interference leads to a different result. Mathematically, two slits make the *amplitude* twice as big at points of constructive interference ($A = 2a$), so the intensity increases by a factor of $2^2 = 4$. Physically, this is conservation of energy. The line labeled $2I_1$ in Figure 22.5a is the uniform intensity that two slits would produce *if* the waves did not interfere. Interference does not change the amount of light energy coming through the two slits, but it does redistribute the light energy on the viewing screen. You can see that the *average* intensity of the

FIGURE 22.5 Intensity of the interference fringes in a double-slit experiment.

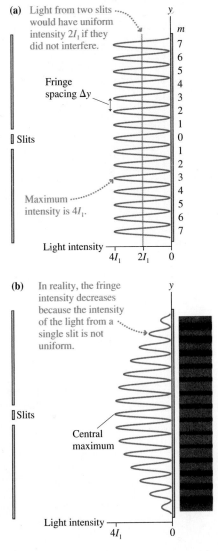

(a) Light from two slits would have uniform intensity $2I_1$ if they did not interfere.

Fringe spacing Δy

Slits

Maximum intensity is $4I_1$.

Light intensity — $4I_1$ $2I_1$ 0

(b) In reality, the fringe intensity decreases because the intensity of the light from a single slit is not uniform.

Slits

Central maximum

Light intensity — $4I_1$ 0

oscillating curve is $2I_1$, but the intensity of the bright fringes gets pushed up from $2I_1$ to $4I_1$ in order for the intensity of the dark fringes to drop from $2I_1$ to 0.

There is still one problem. Equation 22.14 predicts that all interference fringes are equally bright, but you saw in Figure 22.3b that the fringes decrease in brightness as you move away from the center. The erroneous prediction stems from our assumption that the amplitude a of the wave from each slit is constant across the screen. This isn't really true. A more detailed calculation, in which the amplitude gradually decreases as you move away from the center, finds that Equation 22.14 is correct if I_1 slowly decreases as y increases.

FIGURE 22.5b summarizes this analysis by graphing the light intensity (Equation 22.14) with I_1 slowly decreasing as y increases. Comparing this graph to the image, you can see that the wave model of light has provided an excellent description of Young's double-slit interference experiment.

22.3 The Diffraction Grating

Suppose we were to replace the double slit with an opaque screen that has N closely spaced slits. When illuminated from one side, each of these slits becomes the source of a light wave that diffracts, or spreads out, behind the slit. Such a multi-slit device is called a **diffraction grating**. The light intensity pattern on a screen behind a diffraction grating is due to the interference of N overlapped waves.

FIGURE 22.6 shows a diffraction grating in which N slits are equally spaced a distance d apart. This is a top view of the grating, as we look down on the experiment, and the slits extend above and below the page. Only 10 slits are shown here, but a practical grating will have hundreds or even thousands of slits. Suppose a plane wave of wavelength λ approaches from the left. The crest of a plane wave arrives *simultaneously* at each of the slits, causing the wave emerging from each slit to be in phase with the wave emerging from every other slit. Each of these emerging waves spreads out, just like the light wave in Figure 22.2, and after a short distance they all overlap with each other and interfere.

We want to know how the interference pattern will appear on a screen behind the grating. The light wave at the screen is the superposition of N waves, from N slits, as they spread and overlap. As we did with the double slit, we'll assume that the distance L to the screen is very large in comparison with the slit spacing d; hence the path followed by the light from one slit to a point on the screen is *very nearly* parallel to the path followed by the light from neighboring slits. The paths cannot be perfectly parallel, of course, or they would never meet to interfere, but the slight deviation from perfect parallelism is too small to notice. You can see in Figure 22.6 that the wave from one slit travels distance $\Delta r = d \sin\theta$ more than the wave from the slit above it and $\Delta r = d \sin\theta$ less than the wave below it. This is the same reasoning we used in Figure 22.4 to analyze the double-slit experiment.

Figure 22.6 is a magnified view of the slits. FIGURE 22.7 steps back to where we can see the viewing screen. If the angle θ is such that $\Delta r = d \sin\theta = m\lambda$, where m is an integer, then the light wave arriving at the screen from one slit will be *exactly in phase* with the light waves arriving from the two slits next to it. But each of those waves is in phase with waves from the slits next to them, and so on until we reach the end of the grating. In other words, N **light waves, from N different slits, will *all* be in phase with each other** when they arrive at a point on the screen at angle θ_m such that

$$d \sin\theta_m = m\lambda \qquad m = 0, 1, 2, 3, \ldots \qquad (22.15)$$

The screen will have bright constructive-interference fringes at the values of θ_m given by Equation 22.15. We say that the light is "diffracted at angle θ_m."

Because it's usually easier to measure distances rather than angles, the position y_m of the mth maximum is

$$y_m = L \tan\theta_m \qquad \text{(positions of bright fringes)} \qquad (22.16)$$

FIGURE 22.6 Top view of a diffraction grating with $N = 10$ slits.

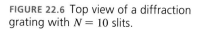

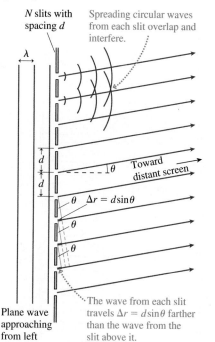

N slits with spacing d

Spreading circular waves from each slit overlap and interfere.

λ

d

d

θ Toward distant screen

$\theta \quad \Delta r = d\sin\theta$

θ

θ

Plane wave approaching from left

The wave from each slit travels $\Delta r = d\sin\theta$ farther than the wave from the slit above it.

FIGURE 22.7 Angles of constructive interference.

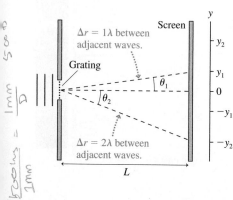

$\Delta r = 1\lambda$ between adjacent waves.

Screen

Grating

θ_1

θ_2

$\Delta r = 2\lambda$ between adjacent waves.

y

y_2

y_1

0

$-y_1$

$-y_2$

L

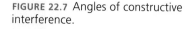

The integer m is called the **order** of the diffraction. For example, light diffracted at θ_2 would be the second-order diffraction. Practical gratings, with very small values for d, display only a few orders. Because d is usually very small, it is customary to characterize a grating by the number of *lines per millimeter*. Here "line" is synonymous with "slit," so the number of lines per millimeter is simply the inverse of the slit spacing d in millimeters.

NOTE ▶ The condition for constructive interference in a grating of N slits is identical to Equation 22.4 for just two slits. Equation 22.15 is simply the requirement that the path-length difference between adjacent slits, be they two or N, is $m\lambda$. But unlike the angles in double-slit interference, the angles of constructive interference from a diffraction grating are generally *not* small angles. The reason is that the slit spacing d in a diffraction grating is so small that λ/d is not a small number. Thus you *cannot* use the small-angle approximation to simplify Equations 22.15 and 22.16. ◀

The wave amplitude at the points of constructive interference is Na because N waves of amplitude a combine in phase. Because the intensity depends on the square of the amplitude, the intensities of the bright fringes of a diffraction grating are

$$I_{max} = N^2 I_1 \tag{22.17}$$

where, as before, I_1 is the intensity of the wave from a single slit. Equation 22.17 is consistent with our prior conclusion that the intensity of a bright fringe in a double-slit interference experiment is four times the intensity of the light from each slit alone. You can see that the fringe intensities increase rapidly as the number of slits increases.

Not only do the fringes get brighter as N increases, they also get narrower. This is again a matter of conservation of energy. If the light waves did not interfere, the intensity from N slits would be NI_1. Interference increases the intensity of the bright fringes by an extra factor of N, so to conserve energy the width of the bright fringes must be proportional to $1/N$. For a realistic diffraction grating, with $N > 100$, the interference pattern consists of a small number of *very* bright and *very* narrow fringes while most of the screen remains dark. FIGURE 22.8a shows the interference pattern behind a diffraction grating both graphically and with a simulation of the viewing screen. A comparison with Figure 22.5b shows that the bright fringes of a diffraction grating are much sharper and more distinct than the fringes of a double slit.

Because the bright fringes are so distinct, diffraction gratings are used for measuring the wavelengths of light. Suppose the incident light consists of two slightly different wavelengths. Each wavelength will be diffracted at a slightly different angle and, if N is sufficiently large, we'll see two distinct fringes on the screen. FIGURE 22.8b illustrates this idea. By contrast, the bright fringes in a double-slit experiment are too broad to distinguish the fringes of one wavelength from those of the other.

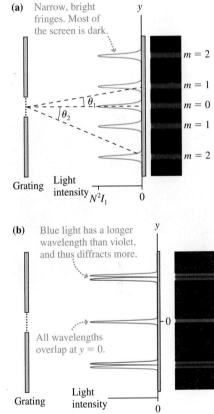

A microscopic side-on look at a diffraction grating.

FIGURE 22.8 The interference pattern behind a diffraction grating.

EXAMPLE 22.3   Measuring wavelengths emitted by sodium atoms

Light from a sodium lamp passes through a diffraction grating having 1000 slits per millimeter. The interference pattern is viewed on a screen 1.000 m behind the grating. Two bright yellow fringes are visible 72.88 cm and 73.00 cm from the central maximum. What are the wavelengths of these two fringes?

VISUALIZE This is the situation shown in Figure 22.8b. The two fringes are very close together, so we expect the wavelengths to be only slightly different. No other yellow fringes are mentioned, so we will assume these two fringes are the first-order diffraction ($m = 1$).

SOLVE The distance y_m of a bright fringe from the central maximum is related to the diffraction angle by $y_m = L\tan\theta_m$. Thus the diffraction angles of these two fringes are

$$\theta_1 = \tan^{-1}\left(\frac{y_1}{L}\right) = \begin{cases} 36.08° & \text{fringe at 72.88 cm} \\ 36.13° & \text{fringe at 73.00 cm} \end{cases}$$

These angles must satisfy the interference condition $d\sin\theta_1 = \lambda$, so the wavelengths are $\lambda = d\sin\theta_1$. What is d? If a 1 mm length of the grating has 1000 slits, then the spacing from one slit to the next must be $1/1000$ mm, or $d = 1.000 \times 10^{-6}$ m. Thus the wavelengths creating the two bright fringes are

$$\lambda = d\sin\theta_1 = \begin{cases} 589.0 \text{ nm} & \text{fringe at 72.88 cm} \\ 589.6 \text{ nm} & \text{fringe at 73.00 cm} \end{cases}$$

ASSESS We had data accurate to four significant figures, and all four were necessary to distinguish the two wavelengths.

The science of measuring the wavelengths of atomic and molecular emissions is called **spectroscopy.** The two sodium wavelengths in this example are called the *sodium doublet,* a name given to two closely spaced wavelengths emitted by the atoms of one element. This doublet is an identifying characteristic of sodium. Because no other element emits these two wavelengths, the doublet can be used to identify the presence of sodium in a sample of unknown composition, even if sodium is only a very minor constituent. This procedure is called *spectral analysis.*

Reflection Gratings

We have analyzed what is called a *transmission grating,* with many parallel slits. In practice, most diffraction gratings are manufactured as *reflection gratings.* The simplest reflection grating, shown in FIGURE 22.9a, is a mirror with hundreds or thousands of narrow, parallel grooves cut into the surface. The grooves divide the surface into many parallel reflective stripes, each of which, when illuminated, becomes the source of a spreading wave. Thus an incident light wave is divided into N overlapped waves. The interference pattern is exactly the same as the interference pattern of light transmitted through N parallel slits.

Naturally occurring reflection gratings are responsible for some forms of color in nature. As the micrograph of FIGURE 22.9b shows, a peacock feather consists of nearly parallel rods of melanin. These act as a reflection grating and create the ever-changing, multicolored hues of iridescence as the angle between the grating and your eye changes. The iridescence of some insects is due to diffraction from parallel microscopic ridges on the shell.

The rainbow of colors reflected from the surface of a DVD is a similar display of interference. The surface of a DVD is smooth plastic with a mirror-like reflective coating in which millions of microscopic holes, each about 1 μm in diameter, encode digital information. From an optical perspective, the array of holes in a shiny surface is a two-dimensional version of the reflection grating shown in Figure 22.9a. Reflection gratings can be manufactured at very low cost simply by stamping holes or grooves into a reflective surface, and these are widely sold as toys and novelty items. Rainbows of color are seen as each wavelength of white light is diffracted at a unique angle.

FIGURE 22.9 Reflection gratings.

(a)

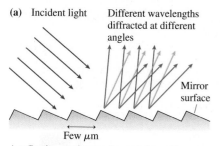

A reflection grating can be made by cutting parallel grooves in a mirror surface. These can be very precise, for scientific use, or mass produced in plastic.

(b)

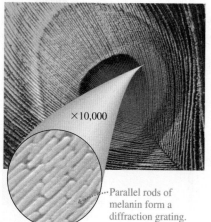

STOP TO THINK 22.3 White light passes through a diffraction grating and forms rainbow patterns on a screen behind the grating. For each rainbow,

a. The red side is on the right, the violet side on the left.
b. The red side is on the left, the violet side on the right.
c. The red side is closest to the center of the screen, the violet side is farthest from the center.
d. The red side is farthest from the center of the screen, the violet side is closest to the center.

22.4 Single-Slit Diffraction

We opened this chapter with a photograph (Figure 22.1a) of a water wave passing through a hole in a barrier, then spreading out on the other side. You then saw an image (Figure 22.2) showing that light, after passing through a very narrow slit, also spreads out on the other side. This phenomenon is called *diffraction.* We're now ready to look at the details of diffraction.

FIGURE 22.10 shows the experimental arrangement for observing the diffraction of light through a narrow slit of width a. Diffraction through a tall, narrow slit is known as **single-slit diffraction.** A viewing screen is placed distance L behind the slit, and we will assume that $L \gg a$. The light pattern on the viewing screen consists of a *central maximum*

FIGURE 22.10 A single-slit diffraction experiment.

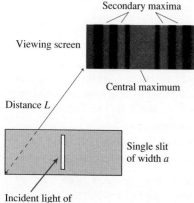

flanked by a series of weaker **secondary maxima** and dark fringes. Notice that the central maximum is significantly broader than the secondary maxima. It is also significantly brighter than the secondary maxima, although that is hard to tell here because this image has been overexposed to make the secondary maxima show up better.

Huygens' Principle

Our analysis of the superposition of waves from distinct sources, such as two loudspeakers or the two slits in a double-slit experiment, has tacitly assumed that the sources are *point sources,* with no measurable extent. To understand diffraction, we need to think about the propagation of an *extended* wave front. This is a problem first considered by the Dutch scientist Christiaan Huygens, a contemporary of Newton who argued that light is a wave.

Huygens lived before a mathematical theory of waves had been developed, so he developed a geometrical model of wave propagation. His idea, which we now call **Huygens' principle,** has two steps:

  1. Each point on a wave front is the source of a spherical *wavelet* that spreads out at the wave speed.
  2. At a later time, the shape of the wave front is the line tangent to all the wavelets.

FIGURE 22.11 illustrates Huygens' principle for a plane wave and a spherical wave. As you can see, the line tangent to the wavelets of a plane wave is a plane that has propagated to the right. The line tangent to the wavelets of a spherical wave is a larger sphere.

Huygens' principle is a visual device, not a theory of waves. Nonetheless, the full mathematical theory of waves, as it developed in the 19th century, justifies Huygens' basic idea, although it is beyond the scope of this textbook to prove it.

Analyzing Single-Slit Diffraction

FIGURE 22.12a shows a wave front passing through a narrow slit of width a. According to Huygens' principle, each point on the wave front can be thought of as the source of a spherical wavelet. These wavelets overlap and interfere, producing the diffraction pattern seen on the viewing screen. The full mathematical analysis, using *every* point on the wave front, is a fairly difficult problem in calculus. We'll be satisfied with a geometrical analysis based on just a few wavelets.

FIGURE 22.12b shows the paths of several wavelets that travel straight ahead to the central point on the screen. (The screen is *very* far to the right in this magnified view of the slit.) The paths are very nearly parallel to each other, thus all the wavelets travel the same distance and arrive at the screen *in phase* with each other. The *constructive interference* between these wavelets produces the central maximum of the diffraction pattern at $\theta = 0$.

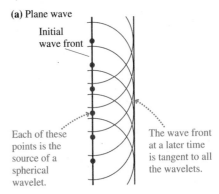

FIGURE 22.11 Huygens' principle applied to the propagation of plane waves and spherical waves.

(a) Plane wave

Initial wave front

Each of these points is the source of a spherical wavelet.

The wave front at a later time is tangent to all the wavelets.

(b) Spherical wave

Initial wave front

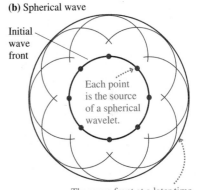

Each point is the source of a spherical wavelet.

The wave front at a later time is tangent to all the wavelets.

FIGURE 22.12 Each point on the wave front is a source of spherical wavelets. The superposition of these wavelets produces the diffraction pattern on the screen.

(a) Greatly magnified view of slit

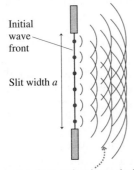

Initial wave front

Slit width a

The wavelets from each point on the initial wave front overlap and interfere, creating a diffraction pattern on the screen.

(b)

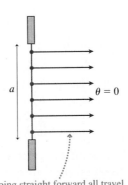

a

$\theta = 0$

The wavelets going straight forward all travel the same distance to the screen. Thus they arrive in phase and interfere constructively to produce the central maximum.

(c)

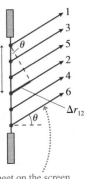

Each point on the wave front is paired with another point distance $a/2$ away.

θ

$\frac{a}{2}$

1
3
5
2
4
6

Δr_{12}

θ

These wavelets all meet on the screen at angle θ. Wavelet 2 travels distance $\Delta r_{12} = (a/2)\sin\theta$ farther than wavelet 1.

The situation is different at points away from the center. Wavelets 1 and 2 in FIGURE 22.12c start from points that are distance $a/2$ apart. If the angle is such that Δr_{12}, the extra distance traveled by wavelet 2, happens to be $\lambda/2$, then wavelets 1 and 2 arrive out of phase and interfere destructively. But if Δr_{12} is $\lambda/2$, then the difference Δr_{34} between paths 3 and 4 and the difference Δr_{56} between paths 5 and 6 are also $\lambda/2$. Those pairs of wavelets also interfere destructively. The superposition of all the wavelets produces perfect destructive interference.

Figure 22.12c shows six wavelets, but our conclusion is valid for any number of wavelets. The key idea is that **every point on the wave front can be paired with another point distance $a/2$ away.** If the path-length difference is $\lambda/2$, the wavelets originating at these two points arrive at the screen out of phase and interfere destructively. When we sum the displacements of all N wavelets, they will—pair by pair—add to zero. The viewing screen at this position will be dark. This is the main idea of the analysis, one worth thinking about carefully.

You can see from Figure 22.12c that $\Delta r_{12} = (a/2)\sin\theta$. This path-length difference will be $\lambda/2$, the condition for destructive interference, if

$$\Delta r_{12} = \frac{a}{2}\sin\theta_1 = \frac{\lambda}{2} \qquad (22.18)$$

or, equivalently, if $a\sin\theta_1 = \lambda$.

> NOTE ▶ Equation 22.18 cannot be satisfied if the slit width a is less than the wavelength λ. If a wave passes through an opening smaller than the wavelength, the central maximum of the diffraction pattern expands to where it *completely* fills the space behind the opening. There are no minima or dark spots at any angle. This situation is uncommon for light waves, because λ is so small, but quite common in the diffraction of sound and water waves. ◀

We can extend this idea to find other angles of perfect destructive interference. Suppose each wavelet is paired with another wavelet from a point $a/4$ away. If Δr between these wavelets is $\lambda/2$, then all N wavelets will again cancel in pairs to give complete destructive interference. The angle θ_2 at which this occurs is found by replacing $a/2$ in Equation 22.18 with $a/4$, leading to the condition $a\sin\theta_2 = 2\lambda$. This process can be continued, and we find that the general condition for complete destructive interference is

$$a\sin\theta_p = p\lambda \qquad p = 1, 2, 3, \ldots \qquad (22.19)$$

When $\theta_p \ll 1$ rad, which is almost always true for light waves, we can use the small-angle approximation to write

$$\theta_p = p\frac{\lambda}{a} \qquad p = 1, 2, 3, \ldots \qquad \text{(angles of dark fringes)} \qquad (22.20)$$

Equation 22.20 gives the angles *in radians* to the dark minima in the diffraction pattern of Figure 22.10. Notice that $p = 0$ is explicitly *excluded*. $p = 0$ corresponds to the straight-ahead position at $\theta = 0$, but you saw in Figures 22.10 and 22.12b that $\theta = 0$ is the central *maximum*, not a minimum.

> NOTE ▶ It is perhaps surprising that Equations 22.19 and 22.20 are *mathematically* the same as the condition for the mth *maximum* of the double-slit interference pattern. But the physical meaning here is quite different. Equation 22.20 locates the *minima* (dark fringes) of the single-slit diffraction pattern. ◀

You might think that we could use this method of pairing wavelets from different points on the wave front to find the maxima in the diffraction pattern. Why not take two points on the wave front that are distance $a/2$ apart, find the angle at which their wavelets are in phase and interfere constructively, then sum over all points on the wave front? There is a subtle but important distinction. FIGURE 22.13 shows six vector

FIGURE 22.13 Destructive interference by pairs leads to net destructive interference, but constructive interference by pairs does *not* necessarily lead to net constructive interference.

(a)

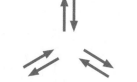

Each pair of vectors interferes destructively. The vector sum of all six vectors is zero.

(b)

Each pair of vectors interferes constructively. Even so, the vector sum of all six vectors is zero.

arrows. The arrows in FIGURE 22.13a are arranged in pairs such that the two members of each pair cancel. The sum of all six vectors is clearly the zero vector $\vec{0}$, representing destructive interference. This is the procedure we used in Figure 22.12c to arrive at Equation 22.18.

The arrows in FIGURE 22.13b are arranged in pairs such that the two members of each pair point in the same direction—constructive interference! Nonetheless, the sum of all six vectors is still $\vec{0}$. To have N waves interfere constructively requires more than simply having constructive interference between pairs. Each pair must also be in phase with every other pair, a condition not satisfied in Figure 22.13b. Constructive interference by pairs does *not* necessarily lead to net constructive interference. It turns out that there is no simple formula to locate the maxima of a single-slit diffraction pattern.

It is possible, although beyond the scope of this textbook, to calculate the entire light intensity pattern. The results of such a calculation are shown graphically in FIGURE 22.14. You can see the bright central maximum at $\theta = 0$, the weaker secondary maxima, and the dark points of destructive interference at the angles given by Equation 22.20. Compare this graph to the image of Figure 22.10 and make sure you see the agreement between the two.

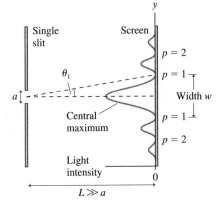

FIGURE 22.14 A graph of the intensity of a single-slit diffraction pattern.

EXAMPLE 22.4 | **Diffraction of a laser through a slit**

Light from a helium-neon laser ($\lambda = 633$ nm) passes through a narrow slit and is seen on a screen 2.0 m behind the slit. The first minimum in the diffraction pattern is 1.2 cm from the central maximum. How wide is the slit?

MODEL A narrow slit produces a single-slit diffraction pattern. A displacement of only 1.2 cm in a distance of 200 cm means that angle θ_1 is certainly a small angle.

VISUALIZE The intensity pattern will look like Figure 22.14.

SOLVE We can use the small-angle approximation to find that the angle to the first minimum is

$$\theta_1 = \frac{1.2 \text{ cm}}{200 \text{ cm}} = 0.00600 \text{ rad} = 0.344°$$

The first minimum is at angle $\theta_1 = \lambda/a$, from which we find that the slit width is

$$a = \frac{\lambda}{\theta_1} = \frac{633 \times 10^{-9} \text{ m}}{6.00 \times 10^{-3} \text{ rad}} = 1.1 \times 10^{-4} \text{ m} = 0.11 \text{ mm}$$

ASSESS This is typical of the slit widths used to observe single-slit diffraction. You can see that the small-angle approximation is well satisfied.

The Width of a Single-Slit Diffraction Pattern

We'll find it useful, as we did for the double slit, to measure positions on the screen rather than angles. The position of the pth dark fringe, at angle θ_p, is $y_p = L\tan\theta_p$, where L is the distance from the slit to the viewing screen. Using Equation 22.20 for θ_p and the small-angle approximation $\tan\theta_p \approx \theta_p$, we find that the dark fringes in the single-slit diffraction pattern are located at

$$y_p = \frac{p\lambda L}{a} \qquad p = 1, 2, 3, \ldots \qquad \text{(positions of dark fringes)} \quad (22.21)$$

A diffraction pattern is dominated by the central maximum, which is much brighter than the secondary maxima. The width w of the central maximum, shown in Figure 22.14, is defined as the distance between the two $p = 1$ minima on either side of the central maximum. Because the pattern is symmetrical, the width is simply $w = 2y_1$. This is

$$w = \frac{2\lambda L}{a} \qquad \text{(single slit)} \qquad (22.22)$$

The width of the central maximum is *twice* the spacing $\lambda L/a$ between the dark fringes on either side. The farther away the screen (larger L), the wider the pattern of light on it becomes. In other words, the light waves are *spreading out* behind the slit, and they fill a wider and wider region as they travel farther.

An important implication of Equation 22.22, one contrary to common sense, is that a narrower slit (smaller a) causes a *wider* diffraction pattern. **The smaller the opening you squeeze a wave through, the *more* it spreads out on the other side.**

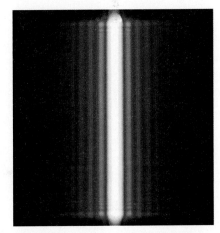

The central maximum of this single-slit diffraction pattern appears white because it is overexposed. The width of the central maximum is clear.

EXAMPLE 22.5 **Determining the wavelength**

Light passes through a 0.12-mm-wide slit and forms a diffraction pattern on a screen 1.00 m behind the slit. The width of the central maximum is 0.85 cm. What is the wavelength of the light?

SOLVE From Equation 22.22, the wavelength is

$$\lambda = \frac{aw}{2L} = \frac{(1.2 \times 10^{-4} \text{ m})(0.0085 \text{ m})}{2(1.00 \text{ m})}$$
$$= 5.1 \times 10^{-7} \text{ m} = 510 \text{ nm}$$

STOP TO THINK 22.4 The figure shows two single-slit diffraction patterns. The distance between the slit and the viewing screen is the same in both cases. Which of the following (perhaps more than one) could be true?

a. The slits are the same for both; $\lambda_1 > \lambda_2$.
b. The slits are the same for both; $\lambda_2 > \lambda_1$.
c. The wavelengths are the same for both; $a_1 > a_2$.
d. The wavelengths are the same for both; $a_2 > a_1$.
e. The slits and the wavelengths are the same for both; $p_1 > p_2$.
f. The slits and the wavelengths are the same for both; $p_2 > p_1$.

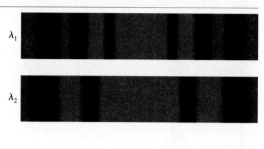

λ_1

λ_2

22.5 Circular-Aperture Diffraction

Diffraction occurs if a wave passes through an opening of any shape. Diffraction by a single slit establishes the basic ideas of diffraction, but a common situation of practical importance is diffraction of a wave by a **circular aperture**. Circular diffraction is mathematically more complex than diffraction from a slit, and we will present results without derivation.

Consider some examples. A loudspeaker cone generates sound by the rapid oscillation of a diaphragm, but the sound wave must pass through the circular aperture defined by the outer edge of the speaker cone before it travels into the room beyond. This is diffraction by a circular aperture. Telescopes and microscopes are the reverse. Light waves from outside need to enter the instrument. To do so, they must pass through a circular lens. In fact, the performance limit of optical instruments is determined by the diffraction of the circular openings through which the waves must pass. This is an issue we'll look at in Chapter 24.

FIGURE 22.15 shows a circular aperture of diameter D. Light waves passing through this aperture spread out to generate a *circular* diffraction pattern. You should compare this to Figure 22.10 for a single slit to note the similarities and differences. The diffraction pattern still has a *central maximum,* now circular, and it is surrounded by a series of secondary bright fringes.

FIGURE 22.15 The diffraction of light by a circular opening.

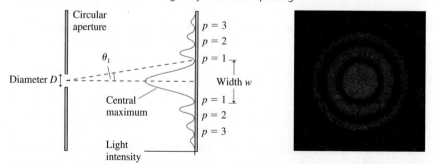

Angle θ_1 locates the first minimum in the intensity, where there is perfect destructive interference. A mathematical analysis of circular diffraction finds

$$\theta_1 = \frac{1.22\lambda}{D} \tag{22.23}$$

where D is the *diameter* of the circular opening. Equation 22.23 has assumed the small-angle approximation, which is almost always valid for the diffraction of light but usually is *not* valid for the diffraction of longer-wavelength sound waves.

Within the small-angle approximation, the width of the central maximum is

$$w = 2y_1 = 2L\tan\theta_1 \approx \frac{2.44\lambda L}{D} \qquad \text{(circular aperture)} \qquad (22.24)$$

The diameter of the diffraction pattern increases with distance L, showing that light spreads out behind a circular aperture, but it decreases if the size D of the aperture is increased.

| EXAMPLE 22.6 | **Shining a laser through a circular hole** |

Light from a helium-neon laser ($\lambda = 633$ nm) passes through a 0.50-mm-diameter hole. How far away should a viewing screen be placed to observe a diffraction pattern whose central maximum is 3.0 mm in diameter?

SOLVE Equation 22.24 gives us the appropriate screen distance:

$$L = \frac{wD}{2.44\lambda} = \frac{(3.0 \times 10^{-3}\text{ m})(5.0 \times 10^{-4}\text{ m})}{2.44(633 \times 10^{-9}\text{ m})} = 0.97\text{ m}$$

The Wave and Ray Models of Light

We opened this chapter by noting that there are three models of light, each useful within a certain range of circumstances. We are now at a point where we can establish an important condition that separates the wave model of light from the ray model of light.

When light passes through an opening of size a, the angle of the first diffraction minimum is

$$\theta_1 = \sin^{-1}\left(\frac{\lambda}{a}\right) \qquad (22.25)$$

Equation 22.25 is for a slit, but the result is very nearly the same if a is the diameter of a circular aperture. Regardless of the shape of the opening, **the factor that determines how much a wave spreads out behind an opening is the ratio λ/a, the size of the wavelength compared to the size of the opening.**

FIGURE 22.16 illustrates the difference between a wave whose wavelength is much smaller than the size of the opening and a second wave whose wavelength is comparable to the opening. A wave with $\lambda/a \approx 1$ quickly spreads to fill the region behind the opening. Light waves, because of their very short wavelength, almost always have $\lambda/a \ll 1$ and diffract to produce a slowly spreading "beam" of light.

Now we can better appreciate Newton's dilemma. With everyday-sized openings, sound and water waves have $\lambda/a \approx 1$ and diffract to fill the space behind the opening. Consequently, this is what we come to expect for the behavior of waves. Newton saw no evidence of this for light passing through openings. We see now that light really does spread out behind an opening, but the very small λ/a ratio usually makes the diffraction pattern too small to see. Diffraction begins to be discernible only when the size of the opening is a fraction of a millimeter or less. If we wanted the diffracted light wave to *fill* the space behind the opening ($\theta_1 \approx 90°$), as a sound wave does, we would need to reduce the size of the opening to $a \approx 0.001$ mm! Although holes this small can be made today, with the processes used to make integrated circuits, the light passing through such a small opening is too weak to be seen by the eye.

FIGURE 22.17 shows light passing through a hole of diameter D. According to the ray model, light rays passing through the hole travel straight ahead to create a bright circular spot of diameter D on a viewing screen. This is the *geometric image* of the slit. In reality, diffraction causes the light to spread out behind the slit, but—and this is the important point—**we will not notice the spreading if it is less than the diameter D of the geometric image.** That is, we will not be aware of diffraction unless the bright spot on the screen increases in diameter.

FIGURE 22.16 The diffraction of a long-wavelength wave and a short-wavelength wave through the same opening.

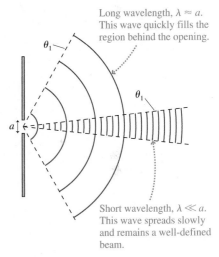

Long wavelength, $\lambda \approx a$. This wave quickly fills the region behind the opening.

Short wavelength, $\lambda \ll a$. This wave spreads slowly and remains a well-defined beam.

FIGURE 22.17 Diffraction will be noticed only if the bright spot on the screen is wider than D.

If light travels in straight lines, the image on the screen is the same size as the hole. Diffraction will not be noticed unless the light spreads over a diameter larger than D.

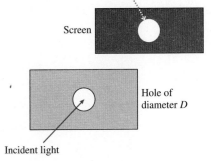

Screen

Hole of diameter D

Incident light

This idea provides a reasonable criterion for when to use ray optics and when to use wave optics:

- If the spreading due to diffraction is less than the size of the opening, use the ray model and think of light as traveling in straight lines.
- If the spreading due to diffraction is greater than the size of the opening, use the wave model of light.

The crossover point between these two regimes occurs when the spreading due to diffraction is equal to the size of the opening. The central-maximum width of a circular-aperture diffraction pattern is $w = 2.44\lambda L/D$. If we equate this diffraction width to the diameter of the aperture itself, we have

$$\frac{2.44\lambda L}{D_c} = D_c \qquad (22.26)$$

where the subscript c on D_c indicates that this is the crossover between the ray model and the wave model. Because we're making an estimate—the change from the ray model to the wave model is gradual, not sudden—to one significant figure, we find

$$D_c \approx \sqrt{2\lambda L} \qquad (22.27)$$

This is the diameter of a circular aperture whose diffraction pattern, at distance L, has width $w = D$. We know that visible light has $\lambda \approx 500$ nm, and a typical distance in laboratory work is $L \approx 1$ m. For these values,

$$D_c \approx 1 \text{ mm}$$

This brings us to an important and very practical conclusion, presented in Tactics Box 22.1.

TACTICS
BOX 22.1   **Choosing a model of light**   (MP)

❶ When visible light passes through openings smaller than about 1 mm in size, diffraction effects are usually important. Use the wave model of light.
❷ When visible light passes through openings larger than about 1 mm in size, diffraction effects are usually not important. Use the ray model of light.

Openings ≈ 1 mm in size are a gray area. Whether one should use a ray model or a wave model will depend on the precise values of λ and L. We'll avoid such ambiguous cases in this book, sticking with examples and homework that fall clearly within the wave model or the ray model. Lenses and mirrors, in particular, are almost always >1 mm in size. We will study the optics of lenses and mirrors in the chapter on ray optics. This chapter on wave optics deals with objects and openings <1 mm in size.

22.6 Interferometers

Scientists and engineers have devised many ingenious methods for using interference to control the flow of light and to make very precise measurements with light waves. A device that makes practical use of interference is called an **interferometer.**

Interference requires two waves of *exactly* the same wavelength. One way of guaranteeing that two waves have exactly equal wavelengths is to divide one wave into two parts of smaller amplitude. Later, at a different point in space, the two parts are recombined. Interferometers are based on the division and recombination of a single wave.

To illustrate the idea, FIGURE 22.18 shows an *acoustical interferometer*. A sound wave is sent into the left end of the tube. The wave splits into two parts at the junction, and waves of smaller amplitude travel around each side. Distance L can be changed by sliding the upper tube in and out like a trombone. After traveling distances r_1 and r_2, the waves recombine and their superposition travels out to the microphone. The sound emerging from the right end has maximum intensity, zero intensity, or somewhere in between depending on the phase difference between the two waves as they recombine.

The two waves traveling through the interferometer started from the *same* source, the loudspeaker; hence the phase difference $\Delta\phi_0$ between the wave sources is automatically zero. The phase difference $\Delta\phi$ between the recombined waves is due entirely to the different distances they travel. Consequently, the conditions for constructive and destructive interference are those we found in Chapter 21 for identical sources:

FIGURE 22.18 An acoustical interferometer.

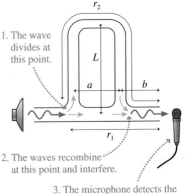

1. The wave divides at this point.

2. The waves recombine at this point and interfere.

3. The microphone detects the superposition of the two waves that traveled different distances.

$$\text{Constructive:}\quad \Delta r = m\lambda$$
$$\text{Destructive:}\quad \Delta r = \left(m + \frac{1}{2}\right)\lambda \qquad m = 0, 1, 2, \ldots \qquad (22.28)$$

The distance each wave travels is easily found from Figure 22.18:

$$r_1 = a + b$$
$$r_2 = L + a + L + b = 2L + a + b$$

Thus the path-length difference between the waves is $\Delta r = r_2 - r_1 = 2L$, and the conditions for constructive and destructive interference are

$$\text{Constructive:}\quad L = m\frac{\lambda}{2}$$
$$\text{Destructive:}\quad L = \left(m + \frac{1}{2}\right)\frac{\lambda}{2} \qquad m = 0, 1, 2, \ldots \qquad (22.29)$$

The interference conditions involve $\lambda/2$ rather than just λ because the wave following the upper path travels distance L *twice*, once up and once down. The upper wave travels a full wavelength λ farther than the lower wave when $L = \lambda/2$.

The interferometer is used by recording the alternating maxima and minima in the sound as the top tube is pulled out and L changes. The interference changes from a maximum to a minimum and back to a maximum every time L increases by half a wavelength. FIGURE 22.19 is a graph of the sound intensity at the microphone as L is increased. You can see, from Equation 22.29, that the number Δm of maxima appearing as the length changes by ΔL is

$$\Delta m = \frac{\Delta L}{\lambda/2} \qquad \lambda = \frac{v_{vac}}{n} \qquad (22.30)$$

Equation 22.30 is the basis for measuring wavelengths very accurately.

FIGURE 22.19 Interference maxima and minima alternate as the slide on an acoustical interferometer is withdrawn.

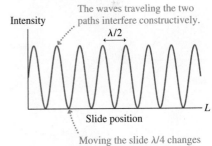

The waves traveling the two paths interfere constructively.

Intensity

$\lambda/2$

Slide position

Moving the slide $\lambda/4$ changes the interference to destructive.

EXAMPLE 22.7 Measuring the wavelength of sound

A loudspeaker broadcasts a sound wave into an acoustical interferometer. The interferometer is adjusted so that the output sound intensity is a maximum, then the slide is slowly withdrawn. Exactly 10 new maxima appear as the slide moves 31.52 cm. What is the wavelength of the sound wave?

MODEL An interferometer produces a new maximum each time L increases by $\lambda/2$, causing the path-length difference Δr to increase by λ.

SOLVE Using Equation 22.30, we have

$$\lambda = \frac{2\Delta L}{\Delta m} = \frac{2(31.52 \text{ cm})}{10} = 6.304 \text{ cm}$$

ASSESS The wavelength can be determined to four significant figures because the distance was measured to four significant figures.

FIGURE 22.20 A Michelson interferometer.

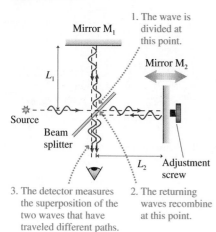

FIGURE 22.20 A Michelson interferometer.

1. The wave is divided at this point.

Mirror M_1

Mirror M_2

L_1

Source

Beam splitter

L_2 Adjustment screw

3. The detector measures the superposition of the two waves that have traveled different paths.

2. The returning waves recombine at this point.

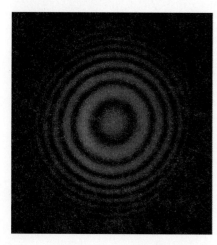

FIGURE 22.21 Photograph of the interference fringes produced by a Michelson interferometer.

The Michelson Interferometer

Albert Michelson, the first American scientist to receive a Nobel Prize, invented an optical interferometer analogous to the acoustical interferometer. In the Michelson interferometer of **FIGURE 22.20**, the light wave is divided by a **beam splitter,** a partially silvered mirror that reflects half the light but transmits the other half. The two waves then travel toward mirrors M_1 and M_2. Half of the wave reflected from M_1 is transmitted through the beam splitter, where it recombines with the reflected half of the wave returning from M_2. The superimposed waves travel on to a light detector, originally a human observer but now more likely an electronic photodetector.

Mirror M_2 can be moved forward or backward by turning a precision screw. This is equivalent to pulling out the slide on the acoustical interferometer. The waves travel distances $r_1 = 2L_1$ and $r_2 = 2L_2$, with the factors of 2 appearing because the waves travel to the mirrors and back again. Thus the path-length difference between the two waves is

$$\Delta r = 2L_2 - 2L_1 \qquad (22.31)$$

The condition for constructive interference is $\Delta r = m\lambda$; hence constructive interference occurs when

$$\text{Constructive:} \qquad L_2 - L_1 = m\frac{\lambda}{2} \qquad m = 0, 1, 2, \ldots \qquad (22.32)$$

This result is essentially identical to Equation 22.29 for an acoustical interferometer. Both divide a wave, send the two smaller waves along two paths that differ in length by Δr, then recombine the two waves at a detector.

You might expect the interferometer output to be either "bright" or "dark." Instead, a viewing screen shows the pattern of circular interference fringes seen in **FIGURE 22.21**. Our analysis was for light waves that impinge on the mirrors exactly perpendicular to the surface. In an actual experiment, some of the light waves enter the interferometer at slightly different angles and, as a result, the recombined waves have slightly altered path-length differences Δr. These waves cause the alternating bright and dark fringes as you move outward from the center of the pattern. Their analysis will be left to more advanced courses in optics. Equation 22.32 is valid at the *center* of the circular pattern; thus there is a bright central spot when Equation 22.32 is true.

If mirror M_2 is moved by turning the screw, the central spot in the fringe pattern alternates between bright and dark. The output recorded by a detector looks exactly like the alternating loud and soft sounds shown in Figure 22.19. Suppose the interferometer is adjusted to produce a bright central spot. The next bright spot will appear when M_2 has moved half a wavelength, increasing the path-length difference by one full wavelength. The number Δm of maxima appearing as M_2 moves through distance ΔL_2 is

$$\Delta m = \frac{\Delta L_2}{\lambda/2} \qquad (22.33)$$

Very precise wavelength measurements can be made by moving the mirror while counting the number of new bright spots appearing at the center of the pattern. The number Δm is counted and known exactly. The only limitation on how precisely λ can be measured this way is the precision with which distance ΔL_2 can be measured. Unlike λ, which is microscopic, ΔL_2 is typically a few millimeters, a macroscopic distance that can be measured very accurately using precision screws, micrometers, and other techniques. Michelson's invention provided a way to transfer the precision of macroscopic distance measurements to an equal precision for the wavelength of light.

EXAMPLE 22.8 **Measuring the wavelength of light**

An experimenter uses a Michelson interferometer to measure one of the wavelengths of light emitted by neon atoms. She slowly moves mirror M_2 until 10,000 new bright central spots have appeared. (In a modern experiment, a photodetector and computer would eliminate the possibility of experimenter error while counting.) She then measures that the mirror has moved a distance of 3.164 mm. What is the wavelength of the light?

MODEL An interferometer produces a new maximum each time L_2 increases by $\lambda/2$.

SOLVE The mirror moves $\Delta L_2 = 3.164$ mm $= 3.164 \times 10^{-3}$ m. We can use Equation 22.33 to find

$$\lambda = \frac{2\Delta L_2}{\Delta m} = 6.328 \times 10^{-7} \text{ m} = 632.8 \text{ nm}$$

ASSESS A measurement of ΔL_2 accurate to four significant figures allowed us to determine λ to four significant figures. This happens to be the neon wavelength that is emitted as the laser beam in a helium-neon laser.

STOP TO THINK 22.5 A Michelson interferometer using light of wavelength λ has been adjusted to produce a bright spot at the center of the interference pattern. Mirror M_1 is then moved distance λ toward the beam splitter while M_2 is moved distance λ away from the beam splitter. How many bright-dark-bright fringe shifts are seen?

a. 0   b. 1   c. 2   d. 4

e. 8   f. It's not possible to say without knowing λ.

Holography

No discussion of wave optics would be complete without mentioning holography, which has both scientific and artistic applications. The basic idea is a simple extension of interferometry.

FIGURE 22.22a shows how a **hologram** is made. A beam splitter divides a laser beam into two waves. One wave illuminates the object of interest. The light scattered by this object is a very complex wave, but it is the wave you would see if you looked at the object from the position of the film. The other wave, called the *reference beam,* is reflected directly toward the film. The scattered light and the reference beam meet at the film and interfere. The film records their interference pattern.

The interference patterns we've looked at in this chapter have been simple patterns of stripes and circles because the light waves have been well-behaved plane waves and spherical waves. The light wave scattered by the object in Figure 22.22a is exceedingly complex. As a result, the interference pattern recorded on the film—the hologram—is a seemingly random pattern of whorls and blotches. FIGURE 22.22b is an enlarged photograph of a portion of a hologram. It's certainly not obvious that information is stored in this pattern, but it is.

A hologram.

FIGURE 22.22 Holography is an important application of wave optics.

(a) Recording a hologram

The interference between the scattered light and the reference beam is recorded on the film.

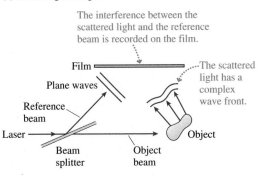

(b) A hologram

An enlarged photo of the developed film. This is the hologram.

(c) Playing a hologram

The diffraction of the laser beam through the light and dark patches of the film reconstructs the original scattered wave.

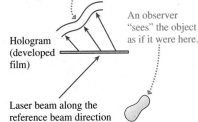

An observer "sees" the object as if it were here.

Hologram (developed film)

Laser beam along the reference beam direction

The hologram is "played" by sending just the reference beam through it, as seen in FIGURE 22.2c. The reference beam diffracts through the transparent parts of the hologram, just as it would through the slits of a diffraction grating. Amazingly, the diffracted wave is *exactly the same* as the light wave that had been scattered by the object! In other words, the diffracted reference beam *reconstructs* the original scattered wave. As you look at this diffracted wave, from the far side of the hologram, you "see" the object exactly as if it were there. The view is three dimensional because, by moving your head with respect to the hologram, you can see different portions of the wave front.

CHALLENGE EXAMPLE 22.9 | **Measuring the index of refraction of a gas**

A Michelson interferometer uses a helium-neon laser with wavelength $\lambda_{vac} = 633$ nm. In one arm, the light passes through a 4.00-cm-thick glass cell. Initially the cell is evacuated, and the interferometer is adjusted so that the central spot is a bright fringe. The cell is then slowly filled to atmospheric pressure with a gas. As the cell fills, 43 bright-dark-bright fringe shifts are seen and counted. What is the index of refraction of the gas at this wavelength?

MODEL Adding one additional wavelength to the round trip causes one bright-dark-bright fringe shift. Changing the length of the arm is one way to add wavelengths, but not the only way. Increasing the index of refraction also adds wavelengths because light has a shorter wavelength when traveling through a material with a larger index of refraction.

VISUALIZE FIGURE 22.23 shows a Michelson interferometer with a cell of thickness d in one arm.

FIGURE 22.23 Measuring the index of refraction.

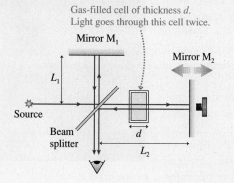

Gas-filled cell of thickness d. Light goes through this cell twice.

Mirror M_1

Mirror M_2

L_1

Source

Beam splitter

d

L_2

SOLVE To begin, all the air is pumped out of the cell. As light travels from the beam splitter to the mirror and back, the number of wavelengths inside the cell is

$$m_1 = \frac{2d}{\lambda_{vac}}$$

where the 2 appears because the light passes through the cell twice.

The cell is then filled with gas at 1 atm pressure. Light travels slower in the gas, $v = c/n$, and you learned in Chapter 20 that the reduction in speed decreases the wavelength to λ_{vac}/n. With the cell filled, the number of wavelengths spanning distance d is

$$m_2 = \frac{2d}{\lambda} = \frac{2d}{\lambda_{vac}/n}$$

The physical distance has not changed, but the number of wavelengths along the path has. Filling the cell has increased the path by

$$\Delta m = m_2 - m_1 = (n-1)\frac{2d}{\lambda_{vac}}$$

wavelengths. Each increase of one wavelength causes one bright-dark-bright fringe shift at the output. Solving for n, we find

$$n = 1 + \frac{\lambda_{vac}\Delta m}{2d} = 1 + \frac{(6.33 \times 10^{-7} \text{ m})(43)}{2(0.0400 \text{ m})} = 1.00034$$

ASSESS This may seem like a six-significant-figure result, but there are really only two. What we're measuring is not n but $n - 1$. We know the fringe count to two significant figures, and that has allowed us to compute $n - 1 = \lambda_{vac}\Delta m/2d = 3.4 \times 10^{-4}$.

SUMMARY

The goal of Chapter 22 has been to understand and apply the wave model of light.

General Principles

Huygens' principle says that each point on a wave front is the source of a spherical wavelet. The wave front at a later time is tangent to all the wavelets.

Diffraction is the spreading of a wave after it passes through an opening.

Constructive and destructive interference are due to the overlap of two or more waves as they spread behind openings.

Important Concepts

The wave model of light considers light to be a wave propagating through space. Diffraction and interference are important.

The ray model of light considers light to travel in straight lines like little particles. Diffraction and interference are not important.

Diffraction is important when the width of the diffraction pattern of an aperture equals or exceeds the size of the aperture. For a circular aperture, the crossover between the ray and wave models occurs for an opening of diameter $D_c \approx \sqrt{2\lambda L}$.

In practice, $D_c \approx 1$ mm for visible light. Thus

- Use the wave model when light passes through openings < 1 mm in size. Diffraction effects are usually important.

- Use the ray model when light passes through openings > 1 mm in size. Diffraction is usually not important.

Applications

Single slit of width a.
A bright **central maximum** of width

$$w = \frac{2\lambda L}{a}$$

is flanked by weaker **secondary maxima**.
Dark fringes are located at angles such that

$$a \sin \theta_p = p\lambda \qquad p = 1, 2, 3, \ldots$$

If $\lambda/a \ll 1$, then from the small-angle approximation

$$\theta_p = \frac{p\lambda}{a} \qquad y_p = \frac{p\lambda L}{a}$$

Circular aperture of diameter D.
A bright central maximum of diameter

$$w = \frac{2.44\lambda L}{D}$$

is surrounded by circular secondary maxima.
The first dark fringe is located at

$$\theta_1 = \frac{1.22\lambda}{D} \qquad y_1 = \frac{1.22\lambda L}{D}$$

For an aperture of any shape, a smaller opening causes a more rapid spreading of the wave behind the opening.

Interference due to wave-front division

Waves overlap as they spread out behind slits. Constructive interference occurs along antinodal lines. Bright fringes are seen where the antinodal lines intersect the viewing screen.

Double slit with separation d.
Equally spaced bright fringes are located at

$$\theta_m = \frac{m\lambda}{d} \qquad y_m = \frac{m\lambda L}{d} \qquad m = 0, 1, 2, \ldots$$

The **fringe spacing** is $\Delta y = \dfrac{\lambda L}{d}$

Diffraction grating with slit spacing d.
Very bright and narrow fringes are located at angles and positions

$$d \sin \theta_m = m\lambda \qquad y_m = L \tan \theta_m$$

Interference due to amplitude division

An interferometer divides a wave, lets the two waves travel different paths, then recombines them. Interference is constructive if one wave travels an integer number of wavelengths more or less than the other wave. The difference can be due to an actual path-length difference or to a different index of refraction.

Michelson interferometer

The number of bright-dark-bright fringe shifts as mirror M_2 moves distance ΔL_2 is

$$\Delta m = \frac{\Delta L_2}{\lambda/2}$$

Terms and Notation

optics	photon model	diffraction grating	Huygens' principle
diffraction	double slit	order, m	circular aperture
models of light	interference fringes	spectroscopy	interferometer
wave model	central maximum	single-slit diffraction	beam splitter
ray model	fringe spacing, Δy	secondary maxima	hologram

CONCEPTUAL QUESTIONS

1. **FIGURE Q22.1** shows light waves passing through two closely spaced, narrow slits. The graph shows the intensity of light on a screen behind the slits. Reproduce these graph axes, including the zero and the tick marks locating the double-slit fringes, then draw a graph to show how the light-intensity pattern will appear if the right slit is blocked, allowing light to go through only the left slit. Explain your reasoning.

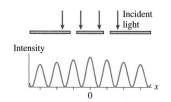

FIGURE Q22.1

2. In a double-slit interference experiment, which of the following actions (perhaps more than one) would cause the fringe spacing to increase? (a) Increasing the wavelength of the light. (b) Increasing the slit spacing. (c) Increasing the distance to the viewing screen. (d) Submerging the entire experiment in water.

3. **FIGURE Q22.3** shows the viewing screen in a double-slit experiment. Fringe C is the central maximum. What will happen to the fringe spacing if
   a. The wavelength of the light is decreased?
   b. The spacing between the slits is decreased?
   c. The distance to the screen is decreased?
   d. Suppose the wavelength of the light is 500 nm. How much farther is it from the dot on the screen in the center of fringe E to the left slit than it is from the dot to the right slit?

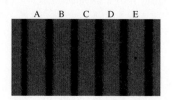

FIGURE Q22.3

4. **FIGURE Q22.3** is the interference pattern seen on a viewing screen behind 2 slits. Suppose the 2 slits were replaced by 20 slits having the same spacing d between adjacent slits.
   a. Would the number of fringes on the screen increase, decrease, or stay the same?
   b. Would the fringe spacing increase, decrease, or stay the same?
   c. Would the width of each fringe increase, decrease, or stay the same?
   d. Would the brightness of each fringe increase, decrease, or stay the same?

5. **FIGURE Q22.5** shows the light intensity on a viewing screen behind a single slit of width a. The light's wavelength is λ. Is $\lambda < a$, $\lambda = a$, $\lambda > a$, or is it not possible to tell? Explain.

FIGURE Q22.5   **FIGURE Q22.6**

6. **FIGURE Q22.6** shows the light intensity on a viewing screen behind a circular aperture. What happens to the width of the central maximum if
   a. The wavelength of the light is increased?
   b. The diameter of the aperture is increased?
   c. How will the screen appear if the aperture diameter is less than the light wavelength?

7. Narrow, bright fringes are observed on a screen behind a diffraction grating. The entire experiment is then immersed in water. Do the fringes on the screen get closer together, get farther apart, remain the same, or disappear? Explain.

8. a. Green light shines through a 100-mm-diameter hole and is observed on a screen. If the hole diameter is increased by 20%, does the circular spot of light on the screen decrease in diameter, increase in diameter, or stay the same? Explain.
   b. Green light shines through a 100-μm-diameter hole and is observed on a screen. If the hole diameter is increased by 20%, does the circular spot of light on the screen decrease in diameter, increase in diameter, or stay the same? Explain.

9. A Michelson interferometer using 800 nm light is adjusted to have a bright central spot. One mirror is then moved 200 nm forward, the other 200 nm back. Afterward, is the central spot bright, dark, or in between? Explain.

10. A Michelson interferometer is set up to display constructive interference (a bright central spot in the fringe pattern of Figure 22.21) using light of wavelength λ. If the wavelength is changed to $\lambda/2$, does the central spot remain bright, does the central spot become dark, or do the fringes disappear? Explain. Assume the fringes are viewed by a detector sensitive to both wavelengths.

EXERCISES AND PROBLEMS

Problems labeled ▨ integrate material from earlier chapters.

Exercises

Section 22.2 The Interference of Light

1. | Two narrow slits 80 μm apart are illuminated with light of wavelength 600 nm. What is the angle of the $m = 3$ bright fringe in radians? In degrees?

2. | A double slit is illuminated simultaneously with orange light of wavelength 600 nm and light of an unknown wavelength. The $m = 4$ bright fringe of the unknown wavelength overlaps the $m = 3$ bright orange fringe. What is the unknown wavelength?

3. | Light of wavelength 500 nm illuminates a double slit, and the interference pattern is observed on a screen. At the position of the $m = 2$ bright fringe, how much farther is it to the more distant slit than to the nearer slit?

4. | A double-slit experiment is performed with light of wavelength 600 nm. The bright interference fringes are spaced 1.8 mm apart on the viewing screen. What will the fringe spacing be if the light is changed to a wavelength of 400 nm?

5. ‖ Light of 600 nm wavelength illuminates a double slit. The intensity pattern shown in FIGURE EX22.5 is seen on a screen 2.0 m behind the slits. What is the spacing (in mm) between the slits?

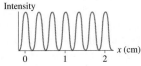

FIGURE EX22.5

6. ‖ Light from a sodium lamp ($\lambda = 589$ nm) illuminates two narrow slits. The fringe spacing on a screen 150 cm behind the slits is 4.0 mm. What is the spacing (in mm) between the two slits?

7. ‖ In a double-slit experiment, the slit separation is 200 times the wavelength of the light. What is the angular separation (in degrees) between two adjacent bright fringes?

8. ‖ A double-slit interference pattern is created by two narrow slits spaced 0.20 mm apart. The distance between the first and the fifth minimum on a screen 60 cm behind the slits is 6.0 mm. What is the wavelength (in nm) of the light used in this experiment?

Section 22.3 The Diffraction Grating

9. | A 4.0-cm-wide diffraction grating has 2000 slits. It is illuminated by light of wavelength 550 nm. What are the angles (in degrees) of the first two diffraction orders?

10. ‖ A diffraction grating produces a first-order maximum at an angle of 20.0°. What is the angle of the second-order maximum?

11. ‖ Light of wavelength 600 nm illuminates a diffraction grating. The second-order maximum is at angle 39.5°. How many lines per millimeter does this grating have?

12. ‖ The two most prominent wavelengths in the light emitted by a hydrogen discharge lamp are 656 nm (red) and 486 nm (blue). Light from a hydrogen lamp illuminates a diffraction grating with 500 lines/mm, and the light is observed on a screen 1.5 m behind the grating. What is the distance between the first-order red and blue fringes?

13. ‖ A helium-neon laser ($\lambda = 633$ nm) illuminates a diffraction grating. The distance between the two $m = 1$ bright fringes is 32 cm on a screen 2.0 m behind the grating. What is the spacing between slits of the grating?

14. | A diffraction grating is illuminated simultaneously with red light of wavelength 660 nm and light of an unknown wavelength. The fifth-order maximum of the unknown wavelength exactly overlaps the third-order maximum of the red light. What is the unknown wavelength?

Section 22.4 Single-Slit Diffraction

15. | A helium-neon laser ($\lambda = 633$ nm) illuminates a single slit and is observed on a screen 1.5 m behind the slit. The distance between the first and second minima in the diffraction pattern is 4.75 mm. What is the width (in mm) of the slit?

16. | In a single-slit experiment, the slit width is 200 times the wavelength of the light. What is the width (in mm) of the central maximum on a screen 2.0 m behind the slit?

17. | The central maximum of a single slit has width 4000λ when viewed on a screen 1.0 m behind the slit. How wide (in mm) is the slit?

18. ‖ Light of 600 nm wavelength illuminates a single slit. The intensity pattern shown in FIGURE EX22.18 is seen on a screen 2.0 m behind the slits. What is the width (in mm) of the slit?

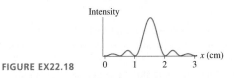

FIGURE EX22.18

19. ‖ A 0.50-mm-wide slit is illuminated by light of wavelength 500 nm. What is the width (in mm) of the central maximum on a screen 2.0 m behind the slit?

20. ‖ You need to use your cell phone, which broadcasts an 800 MHz signal, but you're behind two massive, radio-wave-absorbing buildings that have only a 15 m space between them. What is the angular width, in degrees, of the electromagnetic wave after it emerges from between the buildings?

21. | The opening to a cave is a tall, 30-cm-wide crack. A bat that is preparing to leave the cave emits a 30 kHz ultrasonic chirp. How wide is the "sound beam" 100 m outside the cave opening? Use $v_{\text{sound}} = 340$ m/s.

Section 22.5 Circular-Aperture Diffraction

22. ‖ A 0.50-mm-diameter hole is illuminated by light of wavelength 500 nm. What is the width (in mm) of the central maximum on a screen 2.0 m behind the slit?

23. | Infrared light of wavelength 2.5 μm illuminates a 0.20-mm-diameter hole. What is the angle of the first dark fringe in radians? In degrees?

24. | You want to photograph a circular diffraction pattern whose central maximum has a diameter of 1.0 cm. You have a helium-neon laser ($\lambda = 633$ nm) and a 0.12-mm-diameter pinhole. How far behind the pinhole should you place the screen that's to be photographed?

25. ‖ Light from a helium-neon laser ($\lambda = 633$ nm) passes through a circular aperture and is observed on a screen 4.0 m behind the aperture. The width of the central maximum is 2.5 cm. What is the diameter (in mm) of the hole?

Section 22.6 Interferometers

26. ‖ A Michelson interferometer uses red light with a wavelength of 656.45 nm from a hydrogen discharge lamp. How many bright-dark-bright fringe shifts are observed if mirror M_2 is moved exactly 1 cm?

27. ‖ Moving mirror M_2 of a Michelson interferometer a distance of 100 μm causes 500 bright-dark-bright fringe shifts. What is the wavelength of the light?

28. ‖ A Michelson interferometer uses light whose wavelength is known to be 602.446 nm. Mirror M_2 is slowly moved while exactly 33,198 bright-dark-bright fringe shifts are observed. What distance has M_2 moved? Be sure to give your answer to an appropriate number of significant figures.

29. ‖ A Michelson interferometer uses light from a sodium lamp. Sodium atoms emit light having wavelengths 589.0 nm and 589.6 nm. The interferometer is initially set up with both arms of equal length ($L_1 = L_2$), producing a bright spot at the center of the interference pattern. How far must mirror M_2 be moved so that one wavelength has produced one more new maximum than the other wavelength?

Problems

30. ‖ FIGURE P22.30 shows the light intensity on a screen 2.5 m behind an aperture. The aperture is illuminated with light of wavelength 600 nm.
    a. Is the aperture a single slit or a double slit? Explain.
    b. If the aperture is a single slit, what is its width? If it is a double slit, what is the spacing between the slits?

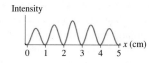

Intensity

FIGURE P22.30   0 1 2 3 4 5 x (cm)

31. ‖ FIGURE P22.31 shows the light intensity on a screen 2.5 m behind an aperture. The aperture is illuminated with light of wavelength 600 nm.
    a. Is the aperture a single slit or a double slit? Explain.
    b. If the aperture is a single slit, what is its width? If it is a double slit, what is the spacing between the slits?

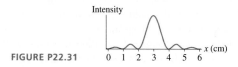

Intensity

FIGURE P22.31   0 1 2 3 4 5 6 x (cm)

32. ‖ Light from a helium-neon laser ($\lambda = 633$ nm) is used to illuminate two narrow slits. The interference pattern is observed on a screen 3.0 m behind the slits. Twelve bright fringes are seen, spanning a distance of 52 mm. What is the spacing (in mm) between the slits?

33. ‖ FIGURE P22.33 shows the light intensity on a screen behind a double slit. The slit spacing is 0.20 mm and the wavelength of the light is 600 nm. What is the distance from the slits to the screen?

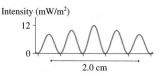

Intensity (mW/m²)

12

0

FIGURE P22.33   2.0 cm

34. ‖ FIGURE P22.33 shows the light intensity on a screen behind a double slit. The slit spacing is 0.20 mm and the screen is 2.0 m behind the slits. What is the wavelength (in nm) of the light?

35. ‖ FIGURE P22.33 shows the light intensity on a screen behind a double slit. Suppose one slit is covered. What will be the light intensity at the center of the screen due to the remaining slit?

36. ‖ A laser beam with a wavelength of 524 nm is exactly perpendicular to a screen having two narrow slits spaced 0.150 mm apart. Interference fringes, including a central maximum, are observed on a viewing screen 1.00 m away. The direction of the laser beam is then slowly rotated by 1.0° around an axis parallel to the slits until it makes an 89.0° angle with the screen. How far does the central maximum move on the viewing screen?

37. ‖ A double-slit experiment is set up using a helium-neon laser ($\lambda = 633$ nm). Then a very thin piece of glass ($n = 1.50$) is placed over one of the slits. Afterward, the central point on the screen is occupied by what had been the $m = 10$ dark fringe. How thick is the glass?

38. ‖ A diffraction grating having 500 lines/mm diffracts visible light at 30°. What is the light's wavelength?

39. ‖ Helium atoms emit light at several wavelengths. Light from a helium lamp illuminates a diffraction grating and is observed on a screen 50.0 cm behind the grating. The emission at wavelength 501.5 nm creates a first-order bright fringe 21.90 cm from the central maximum. What is the wavelength of the bright fringe that is 31.60 cm from the central maximum?

40. ‖ A triple-slit experiment consists of three narrow slits, equally spaced by distance d and illuminated by light of wavelength λ. Each slit alone produces intensity I_1 on the viewing screen at distance L.
    a. Consider a point on the distant viewing screen such that the path-length difference between any two adjacent slits is λ. What is the intensity at this point?
    b. What is the intensity at a point where the path-length difference between any two adjacent slits is $\lambda/2$?

41. ‖ Because sound is a wave, it's possible to make a diffraction grating for sound from a large board of sound-absorbing material with several parallel slits cut for sound to go through. When 10 kHz sound waves pass through such a grating, listeners 10 m from the grating report "loud spots" 1.4 m on both sides of center. What is the spacing between the slits? Use 340 m/s for the speed of sound.

42. ‖ A diffraction grating with 600 lines/mm is illuminated with light of wavelength 500 nm. A very wide viewing screen is 2.0 m behind the grating.
    a. What is the distance between the two $m = 1$ bright fringes?
    b. How many bright fringes can be seen on the screen?

43. ‖ A 500 line/mm diffraction grating is illuminated by light of wavelength 510 nm. How many bright fringes are seen on a 2.0-m-wide screen located 2.0 m behind the grating?

44. ‖ White light (400–700 nm) incident on a 600 line/mm diffraction grating produces rainbows of diffracted light. What is the width of the first-order rainbow on a screen 2.0 m behind the grating?

45. ‖ For your science fair project you need to design a diffraction grating that will disperse the visible spectrum (400–700 nm) over 30.0° in first order.
    a. How many lines per millimeter does your grating need?
    b. What is the first-order diffraction angle of light from a sodium lamp ($\lambda = 589$ nm)?

46. ‖ FIGURE P22.46 shows the interference pattern on a screen 1.0 m behind an 800 line/mm diffraction grating. What is the wavelength (in nm) of the light?

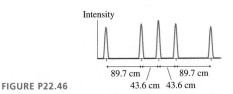

Intensity

89.7 cm ⟋ ⟍ 89.7 cm
43.6 cm 43.6 cm

FIGURE P22.46

47. ‖ FIGURE P22.46 shows the interference pattern on a screen 1.0 m behind a diffraction grating. The wavelength of the light is 600 nm. How many lines per millimeter does the grating have?

48. ‖ Light from a sodium lamp ($\lambda = 589$ nm) illuminates a narrow slit and is observed on a screen 75 cm behind the slit. The distance between the first and third dark fringes is 7.5 mm. What is the width (in mm) of the slit?

49. | The wings of some beetles
BIO have closely spaced parallel lines of melanin, causing the wing to act as a reflection grating. Suppose sunlight shines straight onto a beetle wing. If the melanin lines on the wing are spaced 2.0 μm apart, what is the first-order diffraction angle for green light ($\lambda = 550$ nm)?

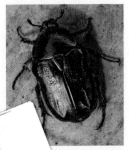

50. | If sunlight shines straight onto a pe⸗⸗⸗⸗er, the feather
BIO appears bright blue when viewed from⸗⸗⸗⸗er side of the incident beam of light. The blue color is⸗⸗⸗⸗raction from parallel rods of melanin in the feather barb⸗⸗s shown in the photograph on page 636. Other wavele⸗⸗ e incident light are diffracted at different angles, leavin⸗⸗ lue light to be seen. The average wavelength of blue⸗⸗70 nm. Assuming this to be the first-order diffract⸗⸗ is the spacing of the melanin rods in the feather?

51. ‖ You've found an unlabeled diffraction gratin⸗⸗ you can use it, you need to know how many lines per ⸗⸗ To find out, you illuminate the grating with light of se⸗⸗ent wavelengths and then measure the distance betw⸗⸗ first-order bright fringes on a viewing screen 150 cm⸗⸗ grating. Your data are as follows:

Wavelength (nm)	Distance (cm)
430	109.6
480	125.4
530	139.8
580	157.2
630	174.4
680	194.8

Use the best-fit line of an appropriate graph to determine the number of lines per mm.

52. ‖ A diffraction grating has slit spacing d. Fringes are viewed on a screen at distance L. What wavelength of light produces a first-order fringe on the viewing screen at distance L from the center of the screen?

53. | For what slit-width-to-wavelength ratio does the first minimum of a single-slit diffraction pattern appear at (a) 30°, (b) 60°, and (c) 90°?

54. ‖ Light from a helium-neon laser ($\lambda = 633$ nm) is incident on a single slit. What is the largest slit width for which there are no minima in the diffraction pattern?

55. ‖ FIGURE P22.55 shows the light intensity on a screen behind a single slit. The slit width is 0.20 mm and the screen is 1.5 m behind the slit. What is the wavelength (in nm) of the light?

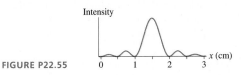

Intensity

FIGURE P22.55

0   1   2   3 x (cm)

56. ‖ FIGURE P22.55 shows the light intensity on a screen behind a single slit. The wavelength of the light is 600 nm and the slit width is 0.15 mm. What is the distance from the slit to the screen?

57. ‖ FIGURE P22.55 shows the light intensity on a screen behind a circular aperture. The wavelength of the light is 500 nm and the screen is 1.0 m behind the slit. What is the diameter (in mm) of the aperture?

58. ‖ Light from a helium-neon laser ($\lambda = 633$ nm) illuminates a circular aperture. It is noted that the diameter of the central maximum on a screen 50 cm behind the aperture matches the diameter of the geometric image. What is the aperture's diameter (in mm)?

59. ‖ One day, after pulling down your window shade, you notice that sunlight is passing through a pinhole in the shade and making a small patch of light on the far wall. Having recently studied optics in your physics class, you're not too surprised to see that the patch of light seems to be a circular diffraction pattern. It appears that the central maximum is about 1 cm across, and you estimate that the distance from the window shade to the wall is about 3 m. Estimate (a) the average wavelength of the sunlight (in nm) and (b) the diameter of the pinhole (in mm).

60. | A radar for tracking aircraft broadcasts a 12 GHz microwave beam from a 2.0-m-diameter circular radar antenna. From a wave perspective, the antenna is a circular aperture through which the microwaves diffract.
    a. What is the diameter of the radar beam at a distance of 30 km?
    b. If the antenna emits 100 kW of power, what is the average microwave intensity at 30 km?

61. ‖ Scientists use *laser range-finding* to measure the distance to the moon with great accuracy. A brief laser pulse is fired at the moon, then the time interval is measured until the "echo" is seen by a telescope. A laser beam spreads out as it travels because it diffracts through a circular exit as it leaves the laser. In order for the reflected light to be bright enough to detect, the laser spot ⸗he moon must be no more than 1.0 km in diameter. Staying ⸗n this diameter is accomplished by using a special large-⸗er laser. If $\lambda = 532$ nm, what is the minimum diameter ⸗rcular opening from which the laser beam emerges? The ⸗on distance is 384,000 km.

62. ‖ Light of wavelength 600 nm passes though two slits separated by 0.20 mm and is observed on a screen 1.0 m behind the slits. The location of the central maximum is marked on the screen and labeled $y = 0$.
    a. At what distance, on either side of $y = 0$, are the $m = 1$ bright fringes?
    b. A very thin piece of glass is then placed in one slit. Because light travels slower in glass than in air, the wave passing through the glass is delayed by 5.0×10^{-16} s in comparison to the wave going through the other slit. What fraction of the period of the light wave is this delay?
    c. With the glass in place, what is the phase difference $\Delta\phi_0$ between the two waves as they leave the slits?
    d. The glass causes the interference fringe pattern on the screen to shift sideways. Which way does the central maximum move (toward or away from the slit with the glass) and by how far?

63. ‖ A 600 line/mm diffraction grating is in an empty aquarium tank. The index of refraction of the glass walls is $n_{glass} = 1.50$. A helium-neon laser ($\lambda = 633$ nm) is outside the aquarium. The laser beam passes through the glass wall and illuminates the diffraction grating.
    a. What is the first-order diffraction angle of the laser beam?
    b. What is the first-order diffraction angle of the laser beam after the aquarium is filled with water ($n_{water} = 1.33$)?

64. | You've set up a Michelson interferometer with a helium-neon laser ($\lambda = 632.8$ nm). After adjusting mirror M_2 to produce a bright spot at the center of the pattern, you carefully move M_2 away from the beam splitter while counting 1200 new bright spots at the center. Then you put the laser away. Later another student wants to restore the interferometer to its starting condition, but he mistakenly sets up a hydrogen discharge lamp and uses the 656.5 nm emission from hydrogen atoms. He then counts 1200 new bright spots while slowly moving M_2 back toward the beam splitter. What is the net displacement of M_2 when he is done? Is M_2 now closer to or farther from the beam splitter?

65. ‖ A Michelson interferometer operating at a 600 nm wavelength has a 2.00-cm-long glass cell in one arm. To begin, the air is pumped out of the cell and mirror M_2 is adjusted to produce a bright spot at the center of the interference pattern. Then a valve is opened and air is slowly admitted into the cell. The index of refraction of air at 1.00 atm pressure is 1.00028. How many bright-dark-bright fringe shifts are observed as the cell fills with air?

66. | A 0.10-mm-thick piece of glass is inserted into one arm of a Michelson interferometer that is using light of wavelength 500 nm. This causes the fringe pattern to shift by 200 fringes. What is the index of refraction of this piece of glass?

67. ‖ Optical computers require microscopic optical switches to turn signals on and off. One device for doing so, which can be implemented in an integrated circuit, is the *Mach-Zender interferometer* seen in FIGURE P22.67. Light from an on-chip infrared laser ($\lambda = 1.000$ μm) is split into two waves that travel equal distances around the arms of the interferometer. One arm passes through an *electro-optic crystal,* a transparent material that can change its index of refraction in response to an applied voltage. Suppose both arms are exactly the same length and the crystal's index of refraction with no applied voltage is 1.522.
    a. With no voltage applied, is the output bright (switch closed, optical signal passing through) or dark (switch open, no signal)? Explain.

b. What is the first index of refraction of the electro-optic crystal larger than 1.522 that changes the optical switch to the state opposite the state you found in part a?

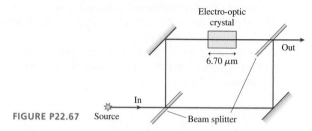

FIGURE P22.67

68. ‖ To illustrate one of the ideas of holography in a simple way, consider a diffraction grating with slit spacing d. The small-angle approximation is usually not valid for diffraction gratings, because d is only slightly larger than λ, but assume that the λ/d ratio of this grating is small enough to make the small-angle approximation valid.
    a. Use the small-angle approximation to find an expression for the fringe spacing on a screen at distance L behind the grating.
    b. Rather than a screen, suppose you place a piece of film at distance L behind the grating. The bright fringes will expose the film, but the dark spaces in between will leave the film unexposed. After being developed, the film will be a series of alternating light and dark stripes. What if you were to now "play" the film by using it as a diffraction grating? In other words, what happens if you shine the same laser through the film and look at the film's diffraction pattern on a screen at the same distance L? Demonstrate that the film's diffraction pattern is a reproduction of the original diffraction grating.

Challenge Problems

69. A helium-neon laser ($\lambda = 633$ nm) is built with a glass tube of inside diameter 1.0 mm, as shown in FIGURE CP22.69. One mirror is partially transmitting to allow the laser beam out. An electrical discharge in the tube causes it to glow like a neon light. From an optical perspective, the laser beam is a light wave that diffracts out through a 1.0-mm-diameter circular opening.
    a. Can a laser beam be *perfectly* parallel, with no spreading? Why or why not?
    b. The angle θ_1 to the first minimum is called the *divergence angle* of a laser beam. What is the divergence angle of this laser beam?
    c. What is the diameter (in mm) of the laser beam after it travels 3.0 m?
    d. What is the diameter of the laser beam after it travels 1.0 km?

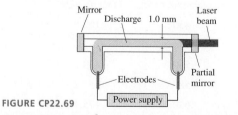

FIGURE CP22.69

70. The intensity at the central maximum of a double-slit interference pattern is $4I_1$. The intensity at the first minimum is zero. At what fraction of the distance from the central maximum to the first minimum is the intensity I_1?

71. Light consisting of two nearly equal wavelengths $\lambda + \Delta\lambda$ and λ, where $\Delta\lambda \ll \lambda$, is incident on a diffraction grating. The slit separation of the grating is d.

   a. Show that the angular separation of these two wavelengths in the mth order is

$$\Delta\theta = \frac{\Delta\lambda}{\sqrt{(d/m)^2 - \lambda^2}}$$

   b. Sodium atoms emit light at 589.0 nm and 589.6 nm. What are the first-order and second-order angular separations (in degrees) of these two wavelengths for a 600 line/mm grating?

72. **FIGURE CP22.72** shows two nearly overlapped intensity peaks of the sort you might produce with a diffraction grating (see Figure 22.8b). As a practical matter, two peaks can just barely be resolved if their spacing Δy equals the width w of each peak, where w is measured at half of the peak's height. Two peaks closer together than w will merge into a single peak. We can use this idea to understand the *resolution* of a diffraction grating.

   a. In the small-angle approximation, the position of the $m = 1$ peak of a diffraction grating falls at the same location as the $m = 1$ fringe of a double slit: $y_1 = \lambda L/d$. Suppose two wavelengths differing by $\Delta\lambda$ pass through a grating at the same time. Find an expression for Δy, the separation of their first-order peaks.

   b. We noted that the widths of the bright fringes are proportional to $1/N$, where N is the number of slits in the grating. Let's hypothesize that the fringe width is $w = y_1/N$. Show that this is true for the double-slit pattern. We'll then assume it to be true as N increases.

   c. Use your results from parts a and b together with the idea that $\Delta y_{min} = w$ to find an expression for $\Delta\lambda_{min}$, the minimum wavelength separation (in first order) for which the diffraction fringes can barely be resolved.

   d. Ordinary hydrogen atoms emit red light with a wavelength of 656.45 nm. In deuterium, which is a "heavy" isotope of hydrogen, the wavelength is 656.27 nm. What is the minimum number of slits in a diffraction grating that can barely resolve these two wavelengths in the first-order diffraction pattern?

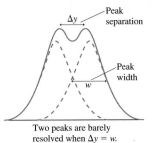

FIGURE CP22.72

73. The diffraction grating analysis in this chapter assumed that the incident light is normal to the grating. **FIGURE CP22.73** shows a plane wave approaching a diffraction grating at angle ϕ.

   a. Show that the angles θ_m for constructive interference are given by the grating equation

$$d(\sin\theta_m + \sin\phi) = m\lambda$$

   where $m = 0, \pm 1, \pm 2, \ldots$. Angles are considered positive if they are above the horizontal line, negative if below it.

   b. The two first-order maxima, $m = +1$ and $m = -1$, are no longer symmetrical about the center. Find θ_1 and θ_{-1} for 500 nm light incident on a 600 line/mm grating at $\phi = 30°$.

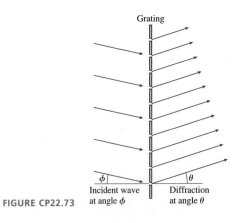

FIGURE CP22.73 — Incident wave at angle ϕ — Diffraction at angle θ

74. **FIGURE CP22.74** shows light of wavelength λ incident at angle ϕ on a *reflection* grating of spacing d. We want to find the angles θ_m at which constructive interference occurs.

   a. The figure shows paths 1 and 2 along which two waves travel and interfere. Find an expression for the path-length difference $\Delta r = r_2 - r_1$.

   b. Using your result from part a, find an equation (analogous to Equation 22.15) for the angles θ_m at which diffraction occurs when the light is incident at angle ϕ. Notice that m can be a negative integer in your expression, indicating that path 2 is shorter than path 1.

   c. Show that the zeroth-order diffraction is simply a "reflection." That is, $\theta_0 = \phi$.

   d. Light of wavelength 500 nm is incident at $\phi = 40°$ on a reflection grating having 700 reflection lines/mm. Find all angles θ_m at which light is diffracted. Negative values of θ_m are interpreted as an angle left of the vertical.

   e. Draw a picture showing a *single* 500 nm light ray incident at $\phi = 40°$ and showing all the diffracted waves at the correct angles.

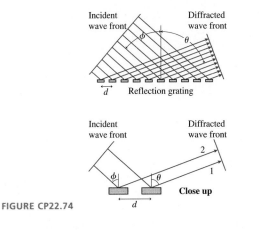

FIGURE CP22.74

75. The pinhole camera of **FIGURE CP22.75** images distant objects by allowing only a narrow bundle of light rays to pass through the hole and strike the film. If light consist-

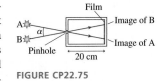

FIGURE CP22.75

ed of particles, you could make the image sharper and sharper (at the expense of getting dimmer and dimmer) by making the aperture smaller and smaller. In practice, diffraction of light by the circular aperture limits the maximum sharpness that can be obtained. Consider two distant points of light, such as two distant streetlights. Each will produce a circular diffraction pattern on the film. The two images can just barely be resolved if the central maximum of one image falls on the first dark fringe of the other image. (This is called Rayleigh's criterion, and we will explore its implication for optical instruments in Chapter 24.)

a. Optimum sharpness of one image occurs when the diameter of the central maximum equals the diameter of the pinhole. What is the optimum hole size for a pinhole camera in which the film is 20 cm behind the hole? Assume $\lambda = 550$ nm, an average value for visible light.
b. For this hole size, what is the angle α (in degrees) between two distant sources that can barely be resolved?
c. What is the distance between two street lights 1 km away that can barely be resolved?

STOP TO THINK ANSWERS

Stop to Think 22.1: b. The antinodal lines seen in Figure 22.3b are diverging.

Stop to Think 22.2: Smaller. Shorter-wavelength light doesn't spread as rapidly as longer-wavelength light. The fringe spacing Δy is directly proportional to the wavelength λ.

Stop to Think 22.3: d. Larger wavelengths have larger diffraction angles. Red light has a larger wavelength than violet light, so red light is diffracted farther from the center.

Stop to Think 22.4: b or c. The width of the central maximum, which is proportional to λ/a, has increased. This could occur either because the wavelength has increased or because the slit width has decreased.

Stop to Think 22.5: d. Moving M_1 in by λ decreases r_1 by 2λ. Moving M_2 out by λ increases r_2 by 2λ. These two actions together change the path length by $\Delta r = 4\lambda$.

23 Ray Optics

The observation that light travels in straight lines—*light rays*—will help us understand the physics of lenses and prisms.

▶ **Looking Ahead** The goals of Chapter 23 are to understand and apply the ray model of light.

The Ray Model of Light

The ray model applies when light interacts with objects that are very large compared to the wavelength. You'll learn that...

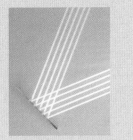

...light rays travel in straight lines unless they are...

...reflected by a surface or...

...refracted at a boundary.

Light rays can also be *scattered* or *absorbed* by the medium they travel through.

Reflection

Light rays can bounce, or **reflect,** off a surface. There are two important cases:

Specular reflection, like from a mirror.

Diffuse reflection, like from the page of this book.

You'll learn to use the *law of reflection.*

Refraction

When light rays travel from one medium to another, they change directions, or **refract,** at the boundary.

Refraction causes the laser beam to change direction as it goes through the prism.

You'll learn to use *Snell's law* to find the angles on both sides.

◀ **Looking Back**
Section 20.5 Index of refraction

Images Formed by Lenses and Mirrors

You'll discover how lenses and mirrors form **images.** We'll start with a graphical method called **ray tracing.**

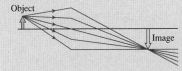

Object

Image

Ray tracing shows how this lens forms a *real image* on the opposite side of the lens from the object.

We'll then develop the **thin-lens equation** for more quantitative results.

A magnifying glass creates a *virtual image* that you see by looking through the lens.

We'll use the same graphical and mathematical techniques to understand how curved mirrors create images.

The passenger-side rearview mirror is curved, allowing you to see a wider field of view.

23.1 The Ray Model of Light

A flashlight makes a beam of light through the night's darkness. Sunbeams stream into a darkened room through a small hole in the shade. Laser beams are even more well defined. Our everyday experience that light travels in straight lines is the basis of the *ray model* of light.

The ray model is an oversimplification of reality but nonetheless is very useful within its range of validity. In particular, the ray model of light is valid as long as any apertures through which the light passes (lenses, mirrors, and holes) are very large compared to the wavelength of light. In that case, diffraction and other wave aspects of light are negligible and can be ignored. The analysis of Section 22.5 found that the crossover between wave optics and ray optics occurs for apertures ≈ 1 mm in diameter. Lenses and mirrors are almost always larger than 1 mm, so the ray model of light is an excellent basis for the practical optics of image formation.

To begin, let us define a **light ray** as a line in the direction along which light energy is flowing. A light ray is an abstract idea, not a physical entity or a "thing." Any narrow beam of light, such as the laser beam in **FIGURE 23.1**, is actually a bundle of many parallel light rays. You can think of a single light ray as the limiting case of a laser beam whose diameter approaches zero. Laser beams are good approximations of light rays, certainly adequate for demonstrating ray behavior, but any real laser beam is a bundle of many parallel rays.

The following table outlines five basic ideas and assumptions of the ray model of light.

FIGURE 23.1 A laser beam or beam of sunlight is a bundle of parallel light rays.

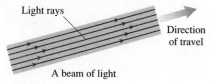

Light rays

Direction of travel

A beam of light

The ray model of light

Light rays travel in straight lines.

Light travels through a transparent material in straight lines called light rays. The speed of light is $v = c/n$, where n is the index of refraction of the material.

Light rays can cross.

Light rays do not interact with each other. Two rays can cross without either being affected in any way.

Material 1   Material 2

Reflection

Refraction

Scattering

Absorption

A light ray travels forever unless it interacts with matter.

A light ray continues forever unless it has an interaction with matter that causes the ray to change direction or to be absorbed. Light interacts with matter in four different ways:

■ At an interface between two materials, light can be either *reflected* or *refracted*.

■ Within a material, light can be either *scattered* or *absorbed*.

These interactions are discussed later in the chapter.

An object is a source of light rays.

An **object** is a source of light rays. Rays originate from *every* point on the object, and each point sends rays in *all* directions. We make no distinction between self-luminous objects and reflective objects.

Diverging bundle of rays

The eye sees by focusing a diverging bundle of rays.

Eye

The eye "sees" an object when *diverging* bundles of rays from each point on the object enter the pupil and are focused to an image on the retina. (Imaging is discussed later in the chapter.) From the movements the eye's lens has to make to focus the image, your brain determines the point from which the rays originated, and you perceive the object as being at that point.

Objects

FIGURE 23.2 illustrates the idea that objects can be either *self-luminous,* such as the sun, flames, and lightbulbs, or *reflective.* Most objects are reflective. A tree, unless it is on fire, is seen or photographed by virtue of reflected sunlight or reflected skylight. People, houses, and this page in the book reflect light from self-luminous sources. In this chapter we are concerned not with how the light originates but with how it behaves after leaving the object.

Light rays from an object are emitted in all directions, but you are not *aware* of light rays unless they enter the pupil of your eye. Consequently, most light rays go completely unnoticed. For example, light rays travel from the sun to the tree in Figure 23.2, but you're not aware of these unless the tree reflects some of them into your eye. Or consider a laser beam. You've probably noticed that it's almost impossible to see a laser beam from the side unless there's dust in the air. The dust scatters a few of the light rays toward your eye, but in the absence of dust you would be completely unaware of a very powerful light beam traveling past you. **Light rays exist independently of whether you are seeing them.**

FIGURE 23.3 shows two idealized sets of light rays. The diverging rays from a **point source** are emitted in all directions. It is useful to think of each point on an object as a point source of light rays. A **parallel bundle** of rays could be a laser beam. Alternatively it could represent a *distant object,* an object such as a star so far away that the rays arriving at the observer are essentially parallel to each other.

Ray Diagrams

Rays originate from *every* point on an object and travel outward in *all* directions, but a diagram trying to show all these rays would be hopelessly messy and confusing. To simplify the picture, we usually use a **ray diagram** showing only a few rays. For example, **FIGURE 23.4** is a ray diagram showing only a few rays leaving the top and bottom points of the object and traveling to the right. These rays will be sufficient to show us how the object is imaged by lenses or mirrors.

NOTE ▶ Ray diagrams are the basis for a *pictorial representation* that we'll use throughout this chapter. Be careful not to think that a ray diagram shows all of the rays. The rays shown on the diagram are just a subset of the infinitely many rays leaving the object. ◀

Apertures

A popular form of entertainment during ancient Roman times was a visit to a **camera obscura,** Latin for "dark room." As **FIGURE 23.5a** shows, a camera obscura was a darkened room with a single, small hole to the outside world. After their eyes became dark adapted, visitors could see a dim but full-color image of the outside world displayed on the back wall of the room. However, the image was upside down! The *pinhole camera* is a miniature version of the camera obscura.

A hole through which light passes is called an **aperture.** **FIGURE 23.5b** uses the ray model of light passing through a small aperture to explain how the camera obscura works. Each point on an object emits light rays in all directions, but only a very few of these rays pass through the aperture and reach the back wall. As the figure illustrates, the geometry of the rays causes the image to be upside down.

Actually, as you may have realized, each *point* on the object illuminates a small but extended *patch* on the wall. This is because the non-zero size of the aperture—needed for the image to be bright enough to see—allows several rays from each point on the object to pass through at slightly different angles. As a result, the image is slightly blurred and out of focus. (Diffraction also becomes an issue if the hole gets too small.) We'll later discover how a modern camera, with a lens, improves on the camera obscura.

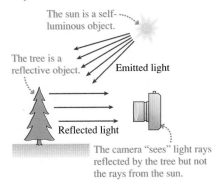

FIGURE 23.2 Self-luminous and reflective objects.

The sun is a self-luminous object.

The tree is a reflective object.

Emitted light

Reflected light

The camera "sees" light rays reflected by the tree but not the rays from the sun.

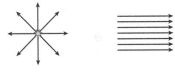

FIGURE 23.3 Point sources and parallel bundles represent idealized objects.

Point source   Parallel bundle

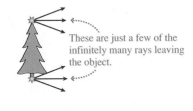

FIGURE 23.4 A ray diagram simplifies the situation by showing only a few rays.

These are just a few of the infinitely many rays leaving the object.

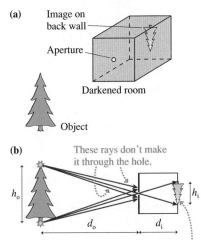

FIGURE 23.5 A camera obscura.

(a)   Image on back wall

Aperture

Darkened room

Object

(b)   These rays don't make it through the hole.

h_o h_i

d_o d_i

The image is upside down. If the hole is sufficiently small, each point on the image corresponds to one point on the object.

FIGURE 23.6 Light through an aperture.

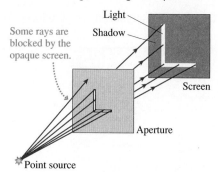

You can see from the similar triangles in Figure 23.5b that the object and image heights are related by

$$\frac{h_i}{h_o} = \frac{d_i}{d_o} \qquad (23.1)$$

where d_o is the distance to the object and d_i is the depth of the camera obscura. Any realistic camera obscura has $d_i < d_o$; thus the image is smaller than the object.

We can apply the ray model to more complex apertures, such as the L-shaped aperture in **FIGURE 23.6**. The pattern of light on the screen is found by tracing all the straight-line paths—the ray trajectories—that start from the point source and pass through the aperture. We will see an enlarged L on the screen, with a sharp boundary between the image and the dark shadow.

STOP TO THINK 23.1 A long, thin lightbulb illuminates a vertical aperture. Which pattern of light do you see on a viewing screen behind the aperture?

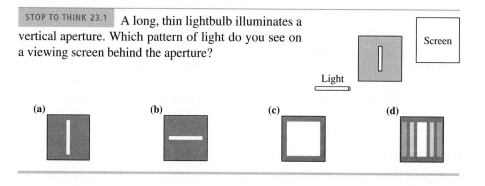

23.2 Reflection

Reflection of light is a familiar, everyday experience. You see your reflection in the bathroom mirror first thing every morning, reflections in your car's rearview mirror as you drive to school, and the sky reflected in puddles of standing water. Reflection from a flat, smooth surface, such as a mirror or a piece of polished metal, is called **specular reflection,** from *speculum,* the Latin word for "mirror."

FIGURE 23.7a shows a bundle of parallel light rays reflecting from a mirror-like surface. You can see that the incident and reflected rays are both in a plane that is normal, or perpendicular, to the reflective surface. A three-dimensional perspective accurately shows the relationship between the light rays and the surface, but figures such as this are hard to draw by hand. Instead, it is customary to represent reflection with the simpler pictorial representation of **FIGURE 23.7b**. In this figure,

FIGURE 23.7 Specular reflection of light.

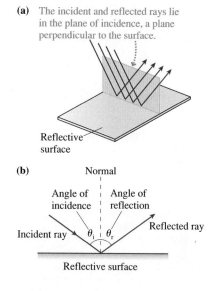

- The plane of the page is the *plane of incidence*, the plane containing both incident and reflected rays. The reflective surface extends into the page.
- A *single* light ray represents the entire bundle of parallel rays. This is oversimplified, but it keeps the figure and the analysis clear.

The angle θ_i between the ray and a line perpendicular to the surface—the *normal* to the surface—is called the **angle of incidence.** Similarly, the **angle of reflection** θ_r is the angle between the reflected ray and the normal to the surface. The **law of reflection,** easily demonstrated with simple experiments, states that

1. The incident ray and the reflected ray are in the same plane normal to the surface, and
2. The angle of reflection equals the angle of incidence: $\theta_r = \theta_i$.

NOTE ▶ Optics calculations *always* use the angle measured from the normal, not the angle between the ray and the surface. ◀

EXAMPLE 23.1 **Light reflecting from a mirror**

A dressing mirror on a closet door is 1.50 m tall. The bottom is 0.50 m above the floor. A bare lightbulb hangs 1.00 m from the closet door, 2.50 m above the floor. How long is the streak of reflected light across the floor?

MODEL Treat the lightbulb as a point source and use the ray model of light.

FIGURE 23.8 Pictorial representation of the light rays reflecting from a mirror.

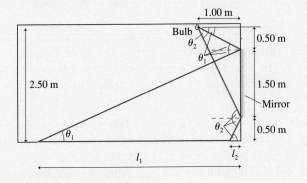

VISUALIZE **FIGURE 23.8** is a pictorial representation of the light rays. We need to consider only the two rays that strike the edges of the mirror. All other reflected rays will fall between these two.

SOLVE Figure 23.8 has used the law of reflection to set the angles of reflection equal to the angles of incidence. Other angles have been identified with simple geometry. The two angles of incidence are

$$\theta_1 = \tan^{-1}\left(\frac{0.50 \text{ m}}{1.00 \text{ m}}\right) = 26.6°$$

$$\theta_2 = \tan^{-1}\left(\frac{2.00 \text{ m}}{1.00 \text{ m}}\right) = 63.4°$$

The distances to the points where the rays strike the floor are then

$$l_1 = \frac{2.00 \text{ m}}{\tan\theta_1} = 4.00 \text{ m}$$

$$l_2 = \frac{0.50 \text{ m}}{\tan\theta_2} = 0.25 \text{ m}$$

Thus the length of the light streak is $l_1 - l_2 = 3.75$ m.

Diffuse Reflection

Most objects are seen by virtue of their reflected light. For a "rough" surface, the law of reflection $\theta_r = \theta_i$ is obeyed at each point but the irregularities of the surface cause the reflected rays to leave in many random directions. This situation, shown in **FIGURE 23.9**, is called **diffuse reflection.** It is how you see this page, the wall, your hand, your friend, and so on.

By a "rough" surface, we mean a surface that is rough or irregular in comparison to the wavelength of light. Because visible-light wavelengths are ≈ 0.5 μm, any surface with texture, scratches, or other irregularities larger than 1 μm will cause diffuse reflection rather than specular reflection. A piece of paper may feel quite smooth to your hand, but a microscope would show that the surface consists of distinct fibers much larger than 1 μm. By contrast, the irregularities on a mirror or a piece of polished metal are much smaller than 1 μm.

FIGURE 23.9 Diffuse reflection from an irregular surface.

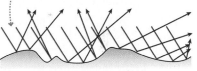

Each ray obeys the law of reflection at that point, but the irregular surface causes the reflected rays to leave in many random directions.

Magnified view of surface

The Plane Mirror

One of the most commonplace observations is that you can see yourself in a mirror. How? **FIGURE 23.10a** shows rays from point source P reflecting from a mirror. Consider the particular ray shown in **FIGURE 23.10b**. The reflected ray travels along a line that passes through point P′ on the "back side" of the mirror. Because $\theta_r = \theta_i$, simple geometry dictates that P′ is the same distance behind the mirror as P is in front of the mirror. That is, $s' = s$.

FIGURE 23.10 The light rays reflecting from a plane mirror.

(a)

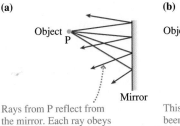

Rays from P reflect from the mirror. Each ray obeys the law of reflection.

(b)

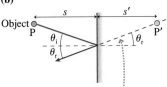

This reflected ray appears to have been traveling along a line that passed through point P′.

(c)

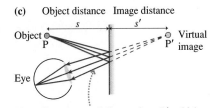

The reflected rays *all* diverge from P′, which appears to be the source of the reflected rays. Your eye collects the bundle of diverging rays and "sees" the light coming from P′.

The location of point P′ in Figure 23.10b is independent of the value of θ_i. Consequently, as FIGURE 23.10c shows, **the reflected rays all *appear* to be coming from point P′**. For a plane mirror, the distance $s′$ to point P′ is equal to the object distance s:

$$s′ = s \quad \text{(plane mirror)} \quad (23.2)$$

If rays diverge from an object point P and interact with a mirror so that the reflected rays diverge from point P′ and *appear* to come from P′, then we call P′ a **virtual image** of point P. The image is "virtual" in the sense that no rays actually leave P′, which is in darkness behind the mirror. But as far as your eye is concerned, the light rays act exactly *as if* the light really originated at P′. So while you may say "I see P in the mirror," what you are actually seeing is the virtual image of P. Distance $s′$ is the *image distance*.

For an extended object, such as the one in FIGURE 23.11, each point on the object from which rays strike the mirror has a corresponding image point an equal distance on the opposite side of the mirror. The eye captures and focuses diverging bundles of rays from each point of the image in order to see the full image in the mirror. Two facts are worth noting:

1. Rays from each point on the object spread out in all directions and strike *every point* on the mirror. Only a very few of these rays enter your eye, but the other rays are very real and might be seen by other observers.
2. Rays from points P and Q enter your eye after reflecting from *different* areas of the mirror. This is why you can't always see the full image of an object in a very small mirror.

FIGURE 23.11 Each point on the extended object has a corresponding image point an equal distance on the opposite side of the mirror.

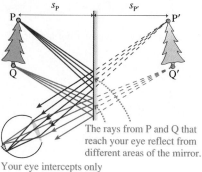

The rays from P and Q that reach your eye reflect from different areas of the mirror.

Your eye intercepts only a very small fraction of all the reflected rays.

EXAMPLE 23.2 **How high is the mirror?**

If your height is h, what is the shortest mirror on the wall in which you can see your full image? Where must the top of the mirror be hung?

MODEL Use the ray model of light.

VISUALIZE FIGURE 23.12 is a pictorial representation of the light rays. We need to consider only the two rays that leave your head and feet and reflect into your eye.

SOLVE Let the distance from your eyes to the top of your head be l_1 and the distance to your feet be l_2. Your height is $h = l_1 + l_2$. A light ray from the top of your head that reflects from the mirror at $\theta_r = \theta_i$ and enters your eye must, by congruent triangles, strike the mirror a distance $\frac{1}{2}l_1$ above your eyes. Similarly, a ray from your foot to your eye strikes the mirror a distance $\frac{1}{2}l_2$ below your eyes. The distance between these two points on the mirror is $\frac{1}{2}l_1 + \frac{1}{2}l_2 = \frac{1}{2}h$. A ray from anywhere else on your body will reach your eye if it strikes the mirror between these two points. Pieces of the mirror outside these two points are irrelevant, not because rays don't strike them but because the reflected rays don't reach your

FIGURE 23.12 Pictorial representation of light rays from your head and feet reflecting into your eye.

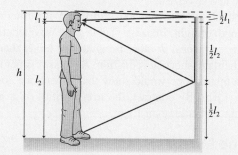

eye. Thus the shortest mirror in which you can see your full reflection is $\frac{1}{2}h$. But this will work only if the top of the mirror is hung midway between your eyes and the top of your head.

ASSESS It is interesting that the answer does not depend on how far you are from the mirror.

STOP TO THINK 23.2 Two plane mirrors form a right angle. How many images of the ball can you see in the mirrors?

a. 1
b. 2
c. 3
d. 4

Observer

23.3 Refraction

Two things happen when a light ray is incident on a smooth boundary between two transparent materials, such as the boundary between air and glass:

1. Part of the light *reflects* from the boundary, obeying the law of reflection. This is how you see reflections from pools of water or storefront windows, even though water and glass are transparent.
2. Part of the light continues into the second medium. It is *transmitted* rather than reflected, but the transmitted ray changes direction as it crosses the boundary. The transmission of light from one medium to another, but with a change in direction, is called **refraction.**

The photograph of FIGURE 23.13 shows the refraction of a light beam as it passes through a glass prism. Notice that the ray direction changes as the light enters and leaves the glass. Our goal in this section is to understand refraction, so we will usually ignore the weak reflection and focus on the transmitted light.

NOTE ▶ A transparent material through which light travels is called the *medium.* This term has to be used with caution. The material does affect the light speed, but a transparent material differs from the medium of a sound or water wave in that particles of the medium do *not* oscillate as a light wave passes through. For a light wave it is the electromagnetic field that oscillates. ◀

FIGURE 23.14a shows the refraction of light rays in a parallel beam of light, such as a laser beam, and rays from a point source. It's good to remember that an infinite number of rays are incident on the boundary, but our analysis will be simplified if we focus on a single light ray. FIGURE 23.14b is a ray diagram showing the refraction of a single ray at a boundary between medium 1 and medium 2. Let the angle between the ray and the normal be θ_1 in medium 1 and θ_2 in medium 2. For the medium in which the ray is approaching the boundary, this is the *angle of incidence* as we've previously defined it. The angle on the transmitted side, *measured from the normal,* is called the **angle of refraction.** Notice that θ_1 is the angle of incidence in Figure 23.14b and the angle of refraction in FIGURE 23.14c, where the ray is traveling in the opposite direction, even though the value of θ_1 has not changed.

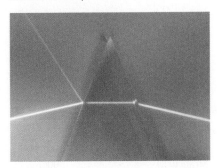

FIGURE 23.13 A light beam refracts twice in passing through a glass prism. You can see a weak reflection from the left surface of the prism.

FIGURE 23.14 Refraction of light rays.

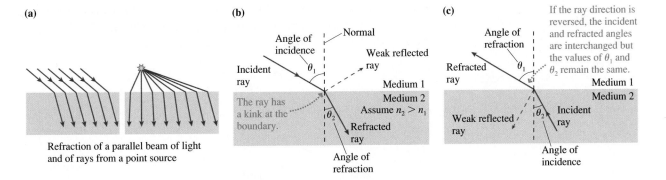

(a) Refraction of a parallel beam of light and of rays from a point source

(b) Angle of incidence, Incident ray, Normal, Weak reflected ray, The ray has a kink at the boundary., θ_1, θ_2, Medium 1, Medium 2, Assume $n_2 > n_1$, Refracted ray, Angle of refraction

(c) Angle of refraction, Refracted ray, θ_1, Weak reflected ray, θ_2, Medium 1, Medium 2, Incident ray, Angle of incidence. If the ray direction is reversed, the incident and refracted angles are interchanged but the values of θ_1 and θ_2 remain the same.

Refraction was first studied experimentally by the Arab scientist Ibn Al-Haitham, in about the year 1000, and later by the Dutch scientist Willebrord Snell. **Snell's law** says that when a ray refracts between medium 1 and medium 2, having indices of refraction n_1 and n_2, the ray angles θ_1 and θ_2 in the two media are related by

$$n_1 \sin\theta_1 = n_2 \sin\theta_2 \qquad \text{(Snell's law of refraction)} \qquad (23.3)$$

Notice that Snell's law does not mention which is the incident angle and which the refracted angle.

The Index of Refraction

To Snell and his contemporaries, n was simply an "index of the refractive power" of a transparent substance. The relationship between the index of refraction and the speed of light was not recognized until the development of a wave theory of light in the 19th century. Theory predicts, and experiment confirms, that light travels through a transparent medium, such as glass or water, at a speed *less* than its speed c in vacuum. In Section 20.5, we defined the *index of refraction n* of a transparent medium as

$$n = \frac{c}{v_{\text{medium}}} \qquad (23.4)$$

where v_{medium} is the light speed in the medium. This implies, of course, that $v_{\text{medium}} = c/n$. The index of refraction of a medium is always $n > 1$ except for vacuum, which has $n = 1$ exactly.

Table 23.1 shows measured values of n for several materials. There are many types of glass, each with a slightly different index of refraction, so we will keep things simple by accepting $n = 1.50$ as a typical value. Notice that cubic zirconia, used to make costume jewelry, has an index of refraction much higher than glass, although not equal to diamond.

We can accept Snell's law as simply an empirical discovery about light. Alternatively, and perhaps surprisingly, we can use the wave model of light to justify Snell's law. The key ideas we need are:

■ Wave fronts represent the crests of waves. They are spaced one wavelength apart.
■ The wavelength in a medium with index of refraction n is $\lambda = \lambda_{\text{vac}}/n$, where λ_{vac} is the vacuum wavelength.
■ Wave fronts are perpendicular to the wave's direction of travel. Consequently, wave fronts are perpendicular to rays.
■ The wave fronts have to stay lined up as a wave crosses from one medium into another.

TABLE 23.1 Indices of refraction

Medium	n
Vacuum	1.00 exactly
Air (actual)	1.0003
Air (accepted)	1.00
Water	1.33
Ethyl alcohol	1.36
Oil	1.46
Glass (typical)	1.50
Polystyrene plastic	1.59
Cubic zirconia	2.18
Diamond	2.41
Silicon (infrared)	3.50

FIGURE 23.15 Snell's law is a consequence of the wave model of light.

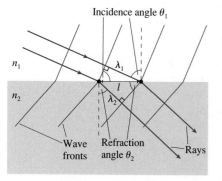

FIGURE 23.15 shows what happens as a wave crosses the boundary between two media, where we're assuming $n_2 > n_1$. **Because the wavelengths differ on opposite sides of the boundary, the wave fronts can stay lined up only if the waves in the two media are traveling in different directions.** In other words, the wave must refract at the boundary to keep the crests of the wave aligned.

To analyze Figure 23.15, consider the segment of boundary of length l between the two dots. This segment is the common hypotenuse of two right triangles. From the upper triangle, which has one side of length λ_1, we see

$$l = \frac{\lambda_1}{\sin\theta_1} \qquad (23.5)$$

where θ_1 is the angle of incidence. Similarly, the lower triangle, where θ_2 is the angle of refraction, gives

$$l = \frac{\lambda_2}{\sin\theta_2} \qquad (23.6)$$

Equating these two expressions for l, and using $\lambda_1 = \lambda_{\text{vac}}/n_1$ and $\lambda_2 = \lambda_{\text{vac}}/n_2$, we find

$$\frac{\lambda_{\text{vac}}}{n_1\sin\theta_1} = \frac{\lambda_{\text{vac}}}{n_2\sin\theta_2} \qquad (23.7)$$

Equation 23.7 can be true only if

$$n_1\sin\theta_1 = n_2\sin\theta_2 \qquad (23.8)$$

which is Snell's law.

Examples of Refraction

Look back at Figure 23.14. As the ray in Figure 23.14b moves from medium 1 to medium 2, where $n_2 > n_1$, it bends closer to the normal. In Figure 23.14c, where the ray moves from medium 2 to medium 1, it bends away from the normal. This is a general conclusion that follows from Snell's law:

- When a ray is transmitted into a material with a higher index of refraction, it bends toward the normal.
- When a ray is transmitted into a material with a lower index of refraction, it bends away from the normal.

This rule becomes a central idea in a procedure for analyzing refraction problems.

TACTICS
BOX 23.1 **Analyzing refraction** (MP)

❶ **Draw a ray diagram.** Represent the light beam with one ray.
❷ **Draw a line normal to the boundary.** Do this at each point where the ray intersects a boundary.
❸ **Show the ray bending in the correct direction.** The angle is larger on the side with the smaller index of refraction. This is the qualitative application of Snell's law.
❹ **Label angles of incidence and refraction.** Measure all angles from the normal.
❺ **Use Snell's law.** Calculate the unknown angle or unknown index of refraction.

Exercises 11–15

EXAMPLE 23.3 **Deflecting a laser beam**

A laser beam is aimed at a 1.0-cm-thick sheet of glass at an angle 30° above the glass.

a. What is the laser beam's direction of travel in the glass?
b. What is its direction in the air on the other side?
c. By what distance is the laser beam displaced?

MODEL Represent the laser beam with a single ray and use the ray model of light.

VISUALIZE **FIGURE 23.16** is a pictorial representation in which the first four steps of Tactics Box 23.1 are identified. Notice that the angle of incidence is $\theta_1 = 60°$, not the 30° value given in the problem.

FIGURE 23.16 The ray diagram of a laser beam passing through a sheet of glass.

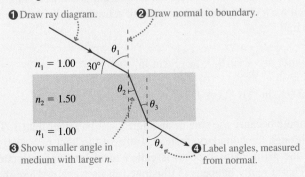

❶ Draw ray diagram. ❷ Draw normal to boundary.

$n_1 = 1.00$ 30° θ_1
$n_2 = 1.50$ θ_2 θ_3
$n_1 = 1.00$ θ_4

❸ Show smaller angle in medium with larger n.
❹ Label angles, measured from normal.

SOLVE a. Snell's law, the final step in the Tactics Box, is $n_1 \sin\theta_1 = n_2 \sin\theta_2$. Using $\theta_1 = 60°$, we find that the direction of travel in the glass is

$$\theta_2 = \sin^{-1}\left(\frac{n_1 \sin\theta_1}{n_2}\right) = \sin^{-1}\left(\frac{\sin 60°}{1.5}\right)$$

$$= \sin^{-1}(0.577) = 35.3°$$

b. Snell's law at the second boundary is $n_2 \sin\theta_3 = n_1 \sin\theta_4$. You can see from Figure 23.16 that the interior angles are equal: $\theta_3 = \theta_2 = 35.3°$. Thus the ray emerges back into the air traveling at angle

$$\theta_4 = \sin^{-1}\left(\frac{n_2 \sin\theta_3}{n_1}\right) = \sin^{-1}(1.5 \sin 35.3°)$$

$$= \sin^{-1}(0.867) = 60°$$

This is the same as θ_1, the original angle of incidence. The glass doesn't change the direction of the laser beam.

c. Although the exiting laser beam is parallel to the initial laser beam, it has been displaced sideways by distance d. **FIGURE 23.17** on the next page shows the geometry for finding d. From trigonometry, $d = l \sin\phi$. Further, $\phi = \theta_1 - \theta_2$ and $l = t/\cos\theta_2$, where t is the thickness of the glass. Combining these gives

Continued

$$d = l\sin\phi = \frac{t}{\cos\theta_2}\sin(\theta_1 - \theta_2)$$

$$= \frac{(1.0\text{ cm})\sin 24.7°}{\cos 35.3°} = 0.51\text{ cm}$$

The glass causes the laser beam to be displaced sideways by 0.51 cm.

ASSESS The laser beam exits the glass still traveling in the same direction as it entered. This is a general result for light traveling through a medium with parallel sides. Notice that the displacement d becomes zero in the limit $t \to 0$. This will be an important observation when we get to lenses.

FIGURE 23.17 The laser beam is deflected sideways by distance d.

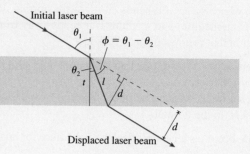

EXAMPLE 23.4 **Measuring the index of refraction**

FIGURE 23.18 shows a laser beam deflected by a 30°-60°-90° prism. What is the prism's index of refraction?

FIGURE 23.18 A prism deflects a laser beam.

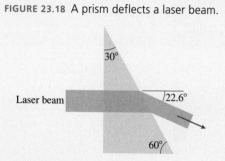

FIGURE 23.19 Pictorial representation of a laser beam passing through the prism.

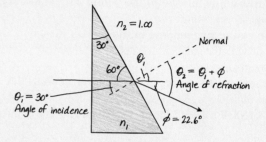

θ_1 and θ_2 are measured from the normal.

MODEL Represent the laser beam with a single ray and use the ray model of light.

VISUALIZE **FIGURE 23.19** uses the steps of Tactics Box 23.1 to draw a ray diagram. The ray is incident perpendicular to the front face of the prism ($\theta_{\text{incident}} = 0°$), thus it is transmitted through the first boundary without deflection. At the second boundary it is especially important to *draw the normal to the surface* at the point of incidence and to *measure angles from the normal*.

SOLVE From the geometry of the triangle you can find that the laser's angle of incidence on the hypotenuse of the prism is

$\theta_1 = 30°$, the same as the apex angle of the prism. The ray exits the prism at angle θ_2 such that the deflection is $\phi = \theta_2 - \theta_1 = 22.6°$. Thus $\theta_2 = 52.6°$. Knowing both angles and $n_2 = 1.00$ for air, we can use Snell's law to find n_1:

$$n_1 = \frac{n_2\sin\theta_2}{\sin\theta_1} = \frac{1.00\sin 52.6°}{\sin 30°} = 1.59$$

ASSESS Referring to the indices of refraction in Table 23.1, we see that the prism is made of plastic.

Total Internal Reflection

FIGURE 23.20 The blue laser beam undergoes total internal reflection inside the prism.

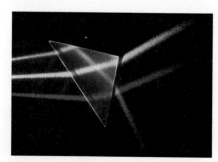

What would have happened in Example 23.4 if the prism angle had been 45° rather than 30°? The light rays would approach the rear surface of the prism at an angle of incidence $\theta_1 = 45°$. When we try to calculate the angle of refraction at which the ray emerges into the air, we find

$$\sin\theta_2 = \frac{n_1}{n_2}\sin\theta_1 = \frac{1.59}{1.00}\sin 45° = 1.12$$

$$\theta_2 = \sin^{-1}(1.12) = \text{???}$$

Angle θ_2 doesn't compute because the sine of an angle can't be larger than 1. The ray is unable to refract through the boundary. Instead, 100% of the light *reflects* from the boundary back into the prism. This process is called **total internal reflection**, often abbreviated TIR. That it really happens is illustrated in FIGURE 23.20. Here three laser beams enter a prism from the left. The bottom two refract out through the right side of the prism. The blue beam, which is incident on the prism's top face, undergoes total internal reflection and then emerges through the right surface.

FIGURE 23.21 shows several rays leaving a point source in a medium with index of refraction n_1. The medium on the other side of the boundary has $n_2 < n_1$. As we've seen, crossing a boundary into a material with a lower index of refraction causes the ray to bend away from the normal. Two things happen as angle θ_1 increases. First, the refraction angle θ_2 approaches 90°. Second, the fraction of the light energy transmitted decreases while the fraction reflected increases.

A **critical angle** is reached when $\theta_2 = 90°$. Because $\sin 90° = 1$, Snell's law $n_1 \sin \theta_c = n_2 \sin 90°$ gives the critical angle of incidence as

$$\theta_c = \sin^{-1}\left(\frac{n_2}{n_1}\right) \qquad (23.9)$$

The refracted light vanishes at the critical angle and the reflection becomes 100% for any angle $\theta_1 \geq \theta_c$. The critical angle is well defined because of our assumption that $n_2 < n_1$. **There is no critical angle and no total internal reflection if $n_2 > n_1$.**

As a quick example, the critical angle in a typical piece of glass at the glass-air boundary is

$$\theta_{c\ glass} = \sin^{-1}\left(\frac{1.00}{1.50}\right) = 42°$$

The fact that the critical angle is less than 45° has important applications. For example, **FIGURE 23.22** shows a pair of binoculars. The lenses are much farther apart than your eyes, so the light rays need to be brought together before exiting the eyepieces. Rather than using mirrors, which get dirty and require alignment, binoculars use a pair of prisms on each side. Thus the light undergoes two total internal reflections and emerges from the eyepiece. (The actual arrangement is a little more complex than in Figure 23.22, to avoid left-right reversals, but this illustrates the basic idea.)

FIGURE 23.21 Refraction and reflection of rays as the angle of incidence increases.

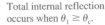
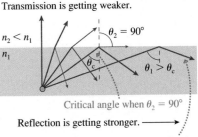

FIGURE 23.22 Binoculars and other optical instruments make use of total internal reflection.

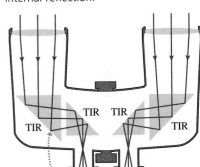

EXAMPLE 23.5 | **Total internal reflection**

A lightbulb is set in the bottom of a 3.0-m-deep swimming pool. What is the diameter of the circle of light seen on the water's surface from above?

MODEL Represent the lightbulb as a point source and use the ray model of light.

VISUALIZE **FIGURE 23.23** is a pictorial representation of the light rays. The lightbulb emits rays at all angles, but only some of the rays refract into the air where they can be seen from above. Rays striking the surface at greater than the critical angle undergo TIR and remain within the water. The diameter of the circle of light is the distance between the two points at which rays strike the surface at the critical angle.

SOLVE From trigonometry, the circle diameter is $D = 2h\tan\theta_c$, where h is the depth of the water. The critical angle for a water-air boundary is $\theta_c = \sin^{-1}(1.00/1.33) = 48.7°$. Thus

$$D = 2(3.0 \text{ m})\tan 48.7° = 6.8 \text{ m}$$

FIGURE 23.23 Pictorial representation of the rays leaving a lightbulb at the bottom of a swimming pool.

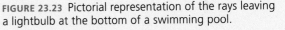
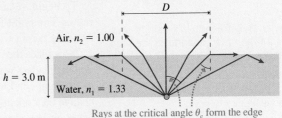

Rays at the critical angle θ_c form the edge of the circle of light seen from above.

Fiber Optics

The most important modern application of total internal reflection is the transmission of light through optical fibers. **FIGURE 23.24a** on the next page shows a laser beam shining into the end of a long, narrow-diameter glass tube. The light rays pass easily from the air into the glass, but they then impinge on the inside wall of the glass tube at an angle

FIGURE 23.24 Light rays are confined within an optical fiber by total internal reflection.

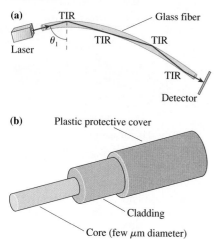

(a)

(b)

of incidence θ_1 approaching 90°. This is well above the critical angle, so the laser beam undergoes TIR and remains inside the glass. The laser beam continues to "bounce" its way down the tube as if the light were inside a pipe. Indeed, optical fibers are sometimes called "light pipes." The rays are *below* the critical angle ($\theta_1 \approx 0$) when they finally reach the end of the fiber, thus they refract out without difficulty and can be detected.

While a simple glass tube can transmit light, reliance on a glass-air boundary is not sufficiently reliable for commercial use. Any small scratch on the side of the tube alters the rays' angle of incidence and allows leakage of light. **FIGURE 23.24b** shows the construction of a practical optical fiber. A small-diameter glass *core* is surrounded by a layer of glass *cladding*. The glasses used for the core and the cladding have $n_{core} > n_{cladding}$; thus light undergoes TIR at the core-cladding boundary and remains confined within the core. This boundary is not exposed to the environment and hence retains its integrity even under adverse conditions.

Even glass of the highest purity is not perfectly transparent. Absorption in the glass, even if very small, causes a gradual decrease in light intensity. The glass used for the core of optical fibers has a minimum absorption at a wavelength of 1.3 μm, in the infrared, so this is the laser wavelength used for long-distance signal transmission. Light at this wavelength can travel hundreds of kilometers through a fiber without significant loss.

STOP TO THINK 23.3 A light ray travels from medium 1 to medium 3 as shown. For these media,

a. $n_3 > n_1$ b. $n_3 = n_1$ c. $n_3 < n_1$
d. We can't compare n_1 to n_3 without knowing n_2.

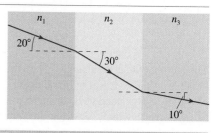

23.4 Image Formation by Refraction

If you see a fish that appears to be swimming close to the front window of the aquarium, but then look through the side of the aquarium, you'll find that the fish is actually farther from the window than you thought. Why is this?

To begin, recall that vision works by focusing a diverging bundle of rays onto the retina. The point from which the rays diverge is where you perceive the object to be. **FIGURE 23.25a** shows how you would see a fish out of water at distance d.

Now place the fish back into the aquarium at the same distance d. For simplicity, we'll ignore the glass wall of the aquarium and consider the water-air boundary. (The thin glass of a typical window has only a very small effect on the refraction of the rays and doesn't change the conclusions.) Light rays again leave the fish, but this time they refract at the water-air boundary. Because they're going from a higher to a lower index of refraction, the rays refract *away from* the normal. **FIGURE 23.25b** shows the consequences.

A bundle of diverging rays still enters your eye, but now these rays are diverging from a closer point, at distance d'. As far as your eye and brain are concerned, it's exactly *as if* the rays really originate at distance d', and this is the location at which you "see" the fish. **The object appears closer than it really is because of the refraction of light at the boundary.**

We found that the rays reflected from a mirror diverge from a point that is not the object point. We called that point a *virtual image*. Similarly, if rays from an object point P refract at a boundary between two media such that the rays then diverge from a point P′ and *appear* to come from P′, we call P′ a virtual image of point P. The virtual image of the fish is what you see.

Let's examine this image formation a bit more carefully. **FIGURE 23.26** shows a boundary between two transparent media having indices of refraction n_1 and n_2. Point P, a source of light rays, is the object. Point P′, from which the rays *appear* to diverge, is

FIGURE 23.25 Refraction of the light rays causes a fish in the aquarium to be seen at distance d'.

(a) A fish out of water

The rays that reach the eye are diverging from this point, the object.

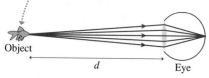

(b) A fish in the aquarium

Refraction causes the rays to bend at the boundary.

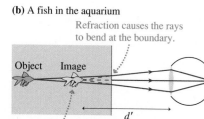

Now the rays that reach the eye are diverging from this point, the image.

the virtual image of P. Distance s is called the **object distance.** Our goal is to determine distance s', the **image distance. Both are measured from the boundary.**

A line perpendicular to the boundary is called the **optical axis.** Consider a ray leaving the object at angle θ_1 with respect to the optical axis. θ_1 is also the angle of incidence at the boundary, where the ray refracts into the second medium at angle θ_2. By tracing the refracted ray backward, you can see that θ_2 is also the angle between the refracted ray and the optical axis at point P$'$.

The distance l is common to both the incident and the refracted rays, and you can see that $l = s \tan \theta_1 = s' \tan \theta_2$. Thus

$$s' = \frac{\tan \theta_1}{\tan \theta_2} s \qquad (23.10)$$

Snell's law relates the sines of angles θ_1 and θ_2; that is,

$$\frac{\sin \theta_1}{\sin \theta_2} = \frac{n_2}{n_1} \qquad (23.11)$$

In practice, the angle between any of these rays and the optical axis is very small because the size of the pupil of your eye is very much less than the distance between the object and your eye. (The angles in the figure have been greatly exaggerated.) Rays that are nearly *parallel* to the *axis* are called **paraxial rays.** The small-angle approximation $\sin \theta \approx \tan \theta \approx \theta$, where θ is in radians, can be applied to paraxial rays. Consequently,

$$\frac{\tan \theta_1}{\tan \theta_2} \approx \frac{\sin \theta_1}{\sin \theta_2} = \frac{n_2}{n_1} \qquad (23.12)$$

Using this result in Equation 23.10, we find that the image distance is

$$s' = \frac{n_2}{n_1} s \qquad (23.13)$$

NOTE ▶ The fact that the result for s' is independent of θ_1 implies that *all* paraxial rays appear to diverge from the same point P$'$. This property of the diverging rays is essential in order to have a well-defined image. ◀

This section has given us a first look at image formation via refraction. We will extend this idea to image formation with lenses in Section 23.6.

FIGURE 23.26 Finding the virtual image P$'$ of an object at P. We've assumed $n_1 > n_2$.

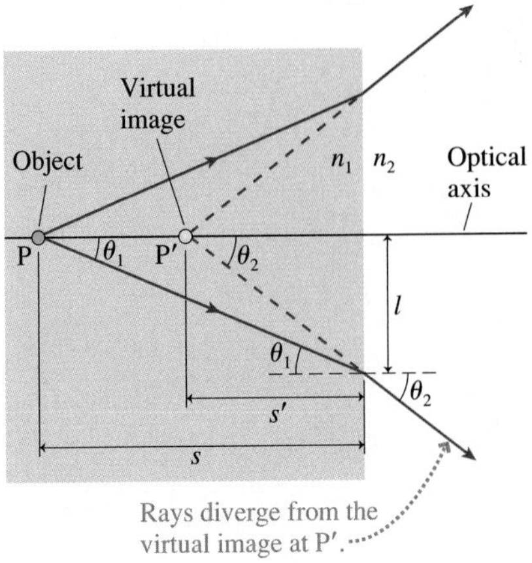

| EXAMPLE 23.6 | **An air bubble in a window** |

A fish and a sailor look at each other through a 5.0-cm-thick glass porthole in a submarine. There happens to be an air bubble right in the center of the glass. How far behind the surface of the glass does the air bubble appear to the fish? To the sailor?

MODEL Represent the air bubble as a point source and use the ray model of light.

VISUALIZE Paraxial light rays from the bubble refract into the air on one side and into the water on the other. The ray diagram looks like Figure 23.26.

SOLVE The index of refraction of the glass is $n_1 = 1.50$. The bubble is in the center of the window, so the object distance from

either side of the window is $s = 2.5$ cm. From the water side, the fish sees the bubble at an image distance

$$s' = \frac{n_2}{n_1} s = \frac{1.33}{1.50}(2.5 \text{ cm}) = 2.2 \text{ cm}$$

This is the apparent depth of the bubble. The sailor, in air, sees the bubble at an image distance

$$s' = \frac{n_2}{n_1} s = \frac{1.00}{1.50}(2.5 \text{ cm}) = 1.7 \text{ cm}$$

ASSESS The image distance is *less* for the sailor because of the *larger* difference between the two indices of refraction.

23.5 Color and Dispersion

One of the most obvious visual aspects of light is the phenomenon of color. Yet color, for all its vivid sensation, is not inherent in the light itself. Color is a *perception*, not a physical quantity. Color is associated with the wavelength of light, but the fact that we see light with a wavelength of 650 nm as "red" tells us how our visual system responds

FIGURE 23.27 Newton used prisms to study color.

(a)

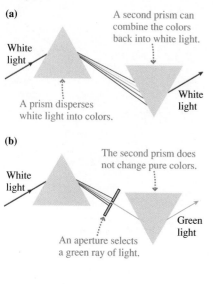

White light

A prism disperses white light into colors.

A second prism can combine the colors back into white light.

White light

(b)

White light

The second prism does not change pure colors.

An aperture selects a green ray of light.

Green light

TABLE 23.2 A brief summary of the visible spectrum of light

Color	Approximate wavelength
Deepest red	700 nm
Red	650 nm
Green	550 nm
Blue	450 nm
Deepest violet	400 nm

FIGURE 23.28 Dispersion curves show how the index of refraction varies with wavelength.

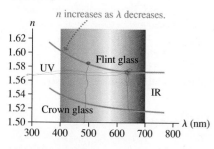

n increases as λ decreases.

to electromagnetic waves of this wavelength. There is no "redness" associated with the light wave itself.

Most of the results of optics do not depend on color. We generally don't need to know the color of light—or, to be more precise, its wavelength—to use the laws of reflection and refraction. Nonetheless, color is an interesting subject, one worthy of a short digression.

Color

It has been known since antiquity that irregularly shaped glass and crystals cause sunlight to be broken into various colors. A common idea was that the glass or crystal somehow altered the properties of the light by *adding* color to the light. Newton suggested a different explanation. He first passed a sunbeam through a prism, producing the familiar rainbow of light. We say that the prism *disperses* the light. Newton's novel idea, shown in FIGURE 23.27a, was to use a second prism, inverted with respect to the first, to "reassemble" the colors. He found that the light emerging from the second prism was a beam of pure, white light.

But the emerging light beam is white only if *all* the rays are allowed to move between the two prisms. Blocking some of the rays with small obstacles, as in FIGURE 23.27b, causes the emerging light beam to have color. This suggests that color is associated with the light itself, not with anything that the prism is "doing" to the light. Newton tested this idea by inserting a small aperture between the prisms to pass only the rays of a particular color, such as green. If the prism alters the properties of light, then the second prism should change the green light to other colors. Instead, the light emerging from the second prism is unchanged from the green light entering the prism.

These and similar experiments show that

1. What we perceive as white light is a mixture of all colors. White light can be dispersed into its various colors and, equally important, mixing all the colors produces white light.
2. The index of refraction of a transparent material differs slightly for different colors of light. Glass has a slightly larger index of refraction for violet light than for green light or red light. Consequently, different colors of light refract at slightly different angles. A prism does not alter the light or add anything to the light; it simply causes the different colors that are inherent in white light to follow slightly different trajectories.

Dispersion

It was Thomas Young, with his two-slit interference experiment, who showed that different colors are associated with light of different wavelengths. The longest wavelengths are perceived as red light and the shortest as violet light. Table 23.2 is a brief summary of the *visible spectrum* of light. Visible-light wavelengths are used so frequently that it is well worth committing this short table to memory.

The slight variation of index of refraction with wavelength is known as **dispersion.** FIGURE 23.28 shows the *dispersion curves* of two common glasses. Notice that **n is *larger* when the wavelength is *shorter*,** thus violet light refracts more than red light.

EXAMPLE 23.7 **Dispersing light with a prism**

Example 23.4 found that a ray incident on a 30° prism is deflected by 22.6° if the prism's index of refraction is 1.59. Suppose this is the index of refraction of deep violet light and deep red light has an index of refraction of 1.54.

a. What is the deflection angle for deep red light?
b. If a beam of white light is dispersed by this prism, how wide is the rainbow spectrum on a screen 2.0 m away?

VISUALIZE Figure 23.19 showed the geometry. A ray of any wavelength is incident on the hypotenuse of the prism at $\theta_1 = 30°$.

SOLVE a. If $n_1 = 1.54$ for deep red light, the refraction angle is

$$\theta_2 = \sin^{-1}\left(\frac{n_1 \sin\theta_1}{n_2}\right) = \sin^{-1}\left(\frac{1.54\sin 30°}{1.00}\right) = 50.4°$$

Example 23.4 showed that the deflection angle is $\phi = \theta_2 - \theta_1$, so deep red light is deflected by $\phi_{red} = 20.4°$. This angle is slightly smaller than the previously observed $\phi_{violet} = 22.6°$.
b. The entire spectrum is spread between $\phi_{red} = 20.4°$ and $\phi_{violet} = 22.6°$. The angular spread is

$$\delta = \phi_{violet} - \phi_{red} = 2.2° = 0.038 \text{ rad}$$

At distance r, the spectrum spans an arc length

$$s = r\delta = (2.0 \text{ m})(0.038 \text{ rad}) = 0.076 \text{ m} = 7.6 \text{ cm}$$

ASSESS The angle is so small that there's no appreciable difference between arc length and a straight line. The spectrum will be 7.6 cm wide at a distance of 2.0 m.

Rainbows

One of the most interesting sources of color in nature is the rainbow. The details get somewhat complicated, but FIGURE 23.29a shows that the basic cause of the rainbow is a combination of refraction, reflection, and dispersion.

Figure 23.29a might lead you to think that the top edge of a rainbow is violet. In fact, the top edge is red, and violet is on the bottom. The rays leaving the drop in Figure 23.29a are spreading apart, so they can't all reach your eye. As FIGURE 23.29b shows, a ray of red light reaching your eye comes from a drop *higher* in the sky than a ray of violet light. In other words, the colors you see in a rainbow refract toward your eye from different raindrops, not from the same drop. You have to look higher in the sky to see the red light than to see the violet light.

FIGURE 23.29 Light seen in a rainbow has undergone refraction + reflection + refraction in a raindrop.

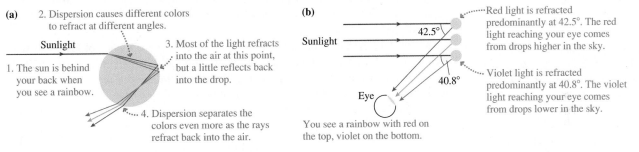

Colored Filters and Colored Objects

White light passing through a piece of green glass emerges as green light. A possible explanation would be that the green glass *adds* "greenness" to the white light, but Newton found otherwise. Green glass is green because it *absorbs* any light that is "not green." We can think of a piece of colored glass or plastic as a *filter* that removes all wavelengths except a chosen few.

If a green filter and a red filter are overlapped, as in FIGURE 23.30, *no* light gets through. The green filter transmits only green light, which is then absorbed by the red filter because it is "not red."

This behavior is true not just for glass filters, which transmit light, but for *pigments* that absorb light of some wavelengths but *reflect* light at other wavelengths. For example, red paint contains pigments reflecting light at wavelengths near 650 nm while absorbing all other wavelengths. Pigments in paints, inks, and natural objects are responsible for most of the color we observe in the world, from the red of lipstick to the blue of a bluebird's feathers.

As an example, FIGURE 23.31 on the next page shows the absorption curve of *chlorophyll*. Chlorophyll is essential for photosynthesis in green plants. The chemical reactions of photosynthesis are able to use red light and blue/violet light, thus chlorophyll absorbs red light and blue/violet light from sunlight and puts it to use. But

FIGURE 23.30 No light at all passes through both a green and a red filter.

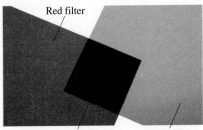

Red filter

Black where filters overlap   Green filter

FIGURE 23.31 The absorption curve of chlorophyll.

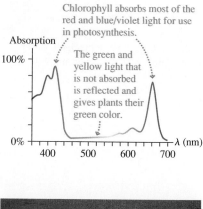

Chlorophyll absorbs most of the red and blue/violet light for use in photosynthesis.

The green and yellow light that is not absorbed is reflected and gives plants their green color.

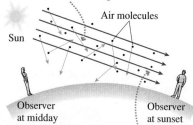

Sunsets are red because all the blue light has scattered as the sunlight passes through the atmosphere.

FIGURE 23.32 Rayleigh scattering by molecules in the air gives the sky and sunsets their color.

At midday the scattered light is mostly blue because molecules preferentially scatter shorter wavelengths.

Air molecules

Sun

Observer at midday

Observer at sunset

At sunset, when the light has traveled much farther through the atmosphere, the light is mostly red because the shorter wavelengths have been lost to scattering.

green and yellow light are not absorbed. Instead, to conserve energy, these wavelengths are mostly *reflected* to give the object a greenish-yellow color. When you look at the green leaves on a tree, you're seeing the light that was reflected because it *wasn't* needed for photosynthesis.

Light Scattering: Blue Skies and Red Sunsets

In the ray model of Section 23.1 we noted that light within a medium can be scattered or absorbed. As we've now seen, the absorption of light can be wavelength dependent and can create color in objects. What are the effects of scattering?

Light can scatter from small particles that are suspended in a medium. If the particles are large compared to the wavelengths of light—even though they may be microscopic and not readily visible to the naked eye—the light essentially reflects off the particles. The law of reflection doesn't depend on wavelength, so all colors are scattered equally. White light scattered from many small particles makes the medium appear cloudy and white. Two well-known examples are clouds, where micrometer-size water droplets scatter the light, and milk, which is a colloidal suspension of microscopic droplets of fats and proteins.

A more interesting aspect of scattering occurs at the atomic level. The atoms and molecules of a transparent medium are much smaller than the wavelengths of light, so they can't scatter light simply by reflection. Instead, the oscillating electric field of the light wave interacts with the electrons in each atom in such a way that the light is scattered. This atomic-level scattering is called **Rayleigh scattering.**

Unlike the scattering by small particles, Rayleigh scattering from atoms and molecules *does* depend on the wavelength. A detailed analysis shows that the intensity of scattered light depends inversely on the fourth power of the wavelength: $I_{\text{scattered}} \propto \lambda^{-4}$. This wavelength dependence explains why the sky is blue and sunsets are red.

As sunlight travels through the atmosphere, the λ^{-4} dependence of Rayleigh scattering causes the shorter wavelengths to be preferentially scattered. If we take 650 nm as a typical wavelength for red light and 450 nm for blue light, the intensity of scattered blue light relative to scattered red light is

$$\frac{I_{\text{blue}}}{I_{\text{red}}} = \left(\frac{650}{450}\right)^4 \approx 4$$

Four times more blue light is scattered toward us than red light and thus, as FIGURE 23.32 shows, the sky appears blue.

Because of the earth's curvature, sunlight has to travel much farther through the atmosphere when we see it at sunrise or sunset than it does during the midday hours. In fact, the path length through the atmosphere at sunset is so long that essentially all the short wavelengths have been lost due to Rayleigh scattering. Only the longer wavelengths remain—orange and red—and they make the colors of the sunset.

23.6 Thin Lenses: Ray Tracing

A camera obscura or a pinhole camera forms images on a screen, but the images are faint and not perfectly focused. The ability to create a bright, well-focused image is vastly improved by using a lens. A **lens** is a transparent material that uses refraction at *curved* surfaces to form an image from diverging light rays. We will defer a mathematical analysis of the refraction of lenses until the next section. First, we want to establish a pictorial method of understanding image formation. This method is called **ray tracing.**

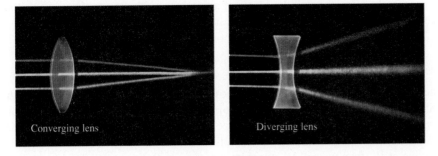

FIGURE 23.33 Parallel light rays pass through a converging lens and a diverging lens.

Converging lens

Diverging lens

FIGURE 23.33 shows parallel light rays entering two different lenses. The left lens, called a **converging lens,** causes the rays to refract *toward* the optical axis. The common point through which initially parallel rays pass is called the **focal point** of the lens. The distance of the focal point from the lens is called the **focal length** f of the lens. The right lens, called a **diverging lens,** refracts parallel rays *away from* the optical axis. This lens also has a focal point, but it is not as obvious.

NOTE ▶ A converging lens is thicker in the center than at the edges. A diverging lens is thicker at the edges than at the center. ◀

FIGURE 23.34 clarifies the situation. In the case of a diverging lens, a backward projection of the diverging rays shows that they *appear* to have started from the same point. This is the focal point of a diverging lens, and its distance from the lens is the focal length of the lens. In the next section we'll relate the focal length to the curvature and index of refraction of the lens, but now we'll use the practical definition that **the focal length is the distance from the lens at which rays parallel to the optical axis converge or from which they diverge.**

NOTE ▶ The focal length f is a property *of the lens,* independent of how the lens is used. The focal length characterizes a lens in much the same way that a mass m characterizes an object or a spring constant k characterizes a spring. ◀

Converging Lenses

These basic observations about lenses are enough to understand image formation by a thin lens. A **thin lens** is a lens whose thickness is very small in comparison to its focal length and in comparison to the object and image distances. We'll make the approximation that the thickness of a thin lens is zero and that the lens lies in a plane called the **lens plane.** Within this approximation, **all refraction occurs as the rays cross the lens plane, and all distances are measured from the lens plane.** Fortunately, the thin-lens approximation is quite good for most practical applications of lenses.

NOTE ▶ We'll *draw* lenses as if they have a thickness, because that is how we expect lenses to look, but our analysis will not depend on the shape or thickness of a lens. ◀

FIGURE 23.35 shows three important situations of light rays passing through a thin converging lens. Part a is familiar from Figure 23.34. If the direction of each of the rays in FIGURE 23.35a is reversed, Snell's law tells us that each ray will exactly retrace its path and emerge from the lens parallel to the optical axis. This leads to FIGURE 23.35b, which is the "mirror image" of part (a). Notice that the lens actually has *two* focal points, located at distances f on either side of the lens.

FIGURE 23.35c shows three rays passing through the *center* of the lens. At the center, the two sides of a lens are very nearly parallel to each other. Earlier, in Example 23.3, we found that a ray passing through a piece of glass with parallel sides is *displaced*

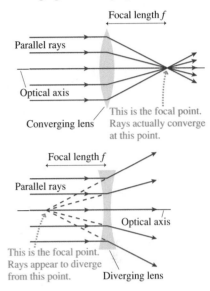

FIGURE 23.34 The focal lengths of converging and diverging lenses.

Focal length f

Parallel rays

Optical axis

Converging lens

This is the focal point. Rays actually converge at this point.

Focal length f

Parallel rays

Optical axis

This is the focal point. Rays appear to diverge from this point.

Diverging lens

FIGURE 23.35 Three important sets of rays passing through a thin converging lens.

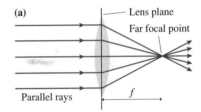

(a)

Lens plane

Far focal point

Parallel rays

f

Any ray initially parallel to the optical axis will refract through the focal point on the far side of the lens.

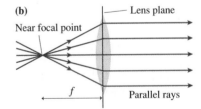

(b)

Lens plane

Near focal point

f

Parallel rays

Any ray passing through the near focal point emerges from the lens parallel to the optical axis.

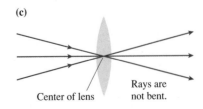

(c)

Center of lens

Rays are not bent.

Any ray directed at the center of the lens passes through in a straight line.

but *not bent* and that the displacement becomes zero as the thickness approaches zero. Consequently, a ray through the center of a thin lens, with zero thickness, is neither bent nor displaced but travels in a straight line.

These three situations form the basis for ray tracing.

Real Images

FIGURE 23.36 shows a lens and an object whose distance from the lens is larger than the focal length. Rays from point P on the object are refracted by the lens so as to converge at point P′ on the opposite side of the lens. If rays diverge from an object point P and interact with a lens such that the refracted rays *converge* at point P′, actually meeting at P′, then we call P′ a **real image** of point P. Contrast this with our prior definition of a *virtual image* as a point from which rays—which never meet—appear to *diverge.*

FIGURE 23.36 Rays from an object point P are refracted by the lens and converge to a real image at point P′.

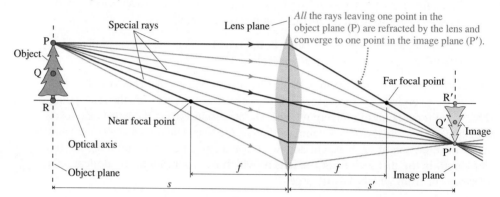

All points on the object that are in the same plane, the **object plane,** converge to image points in the **image plane.** Points Q and R in the object plane of Figure 23.36 have image points Q′ and R′ in the same plane as point P′. Once we locate *one* point in the image plane, such as point P′, we know that the full image lies in the same plane.

There are two important observations to make about Figure 23.36. First, the image is upside down with respect to the object. This is called an **inverted image,** and it is a standard characteristic of real-image formation with a converging lens. Second, rays from point P *fill* the entire lens surface, and all portions of the lens contribute to the image. A larger lens will "collect" more rays and thus make a brighter image.

FIGURE 23.37 is a close-up view of the rays very near the image plane. The rays don't stop at P′ unless we place a screen in the image plane. When we do so, we see a sharp, well-focused image on the screen. To focus an image, you must either move the screen to coincide with the image plane or move the lens or object to make the image plane coincide with the screen. For example, the focus knob on a projector moves the lens forward or backward until the image plane matches the screen position.

NOTE ▶ The ability to see a real image on a screen sets real images apart from *virtual* images. But keep in mind that we need not *see* a real image in order to *have* an image. A real image exists at a point in space where the rays converge even if there's no viewing screen in the image plane. ◀

Figure 23.36 highlights three "special rays" based on the three situations of Figure 23.35. These three rays alone are sufficient to locate the image point P′. That is, we don't need to draw all the rays shown in Figure 23.36. The procedure known as *ray tracing* consists of locating the image by the use of just these three rays.

FIGURE 23.37 A close-up look at the rays near the image plane.

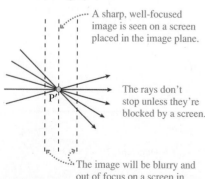

A sharp, well-focused image is seen on a screen placed in the image plane.

The rays don't stop unless they're blocked by a screen.

The image will be blurry and out of focus on a screen in these planes.

TACTICS
BOX 23.2 **Ray tracing for a converging lens** (MP)

❶ **Draw an optical axis.** Use graph paper or a ruler! Establish an appropriate scale.

❷ **Center the lens on the axis.** Mark and label the focal points at distance f on either side.

❸ **Represent the object with an upright arrow at distance s.** It's usually best to place the base of the arrow on the axis and to draw the arrow about half the radius of the lens.

❹ **Draw the three "special rays" from the tip of the arrow.** Use a straight edge.

    a. A ray parallel to the axis refracts through the far focal point.
    b. A ray that enters the lens along a line through the near focal point emerges parallel to the axis.
    c. A ray through the center of the lens does not bend.

❺ **Extend the rays until they converge.** This is the image point. Draw the rest of the image in the image plane. If the base of the object is on the axis, then the base of the image will also be on the axis.

❻ **Measure the image distance s'.** Also, if needed, measure the image height relative to the object height.

Exercises 22–27

EXAMPLE 23.8 **Finding the image of a flower**

A 4.0-cm-diameter flower is 200 cm from the 50-cm-focal-length lens of a camera. How far should the light detector be placed behind the lens to record a well-focused image? What is the diameter of the image on the detector?

MODEL The flower is in the object plane. Use ray tracing to locate the image.

VISUALIZE FIGURE 23.38 shows the ray-tracing diagram and the steps of Tactics Box 23.2. The image has been drawn in the plane where the three special rays converge. You can see *from the drawing* that the image distance is $s' \approx 67$ cm. This is where the detector needs to be placed to record a focused image.

The heights of the object and image are labeled h and h'. The ray through the center of the lens is a straight line, thus the object and image both subtend the same angle θ. Using similar triangles,

$$\frac{h'}{s'} = \frac{h}{s}$$

Solving for h' gives

$$h' = h\frac{s'}{s} = (4.0 \text{ cm})\frac{67 \text{ cm}}{200 \text{ cm}} = 1.3 \text{ cm}$$

The flower's image has a diameter of 1.3 cm.

ASSESS We've been able to learn a great deal about the image from a simple geometric procedure.

FIGURE 23.38 Ray-tracing diagram for Example 23.8.

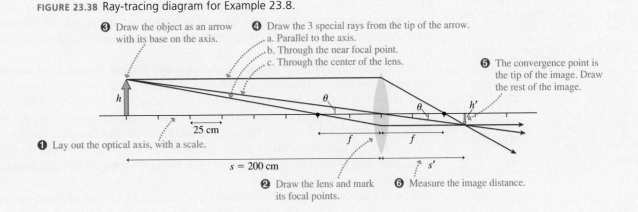

Lateral Magnification

The image can be either larger or smaller than the object, depending on the location and focal length of the lens. But there's more to a description of the image than just its size. We also want to know its *orientation* relative to the object. That is, is the image upright or inverted? It is customary to combine size and orientation information into a single number. The **lateral magnification** *m* is defined as

$$m = -\frac{s'}{s} \tag{23.14}$$

You just saw in Example 23.8 that the image-to-object height ratio is $h'/h = s'/s$. Consequently, we interpret the lateral magnification *m* as follows:

1. A positive value of *m* indicates that the image is upright relative to the object. A negative value of *m* indicates that the image is inverted relative to the object.
2. The absolute value of *m* gives the size ratio of the image and object: $h'/h = |m|$.

The lateral magnification in Example 23.8 would be $m = -0.33$, indicating that the image is inverted and 33% the size of the object.

NOTE ▶ The image-to-object height ratio is called *lateral* magnification to distinguish it from angular magnification, which we'll introduce in the next chapter. In practice, *m* is simply called "magnification" when there's no chance of confusion. Magnification can be less than 1, meaning that the image is smaller than the object. ◀

STOP TO THINK 23.4   A lens produces a sharply focused, inverted image on a screen. What will you see on the screen if the lens is removed?

a. The image will be inverted and blurry.
b. The image will be upright and sharp.
c. The image will be upright and blurry.
d. The image will be much dimmer but otherwise unchanged.
e. There will be no image at all.

Virtual Images

The previous section considered a converging lens with the object at distance $s > f$. That is, the object was outside the focal point. What if the object is inside the focal point, at distance $s < f$? FIGURE 23.39 shows just this situation, and we can use ray tracing to analyze it.

The special rays initially parallel to the axis and through the center of the lens present no difficulties. However, a ray through the near focal point would travel toward the left and would never reach the lens! Referring back to Figure 23.35b, you can see that the rays emerging parallel to the axis entered the lens *along a line* passing through the near focal point. It's the angle of incidence on the lens that is important, not whether the light ray actually passes through the focal point. This was the basis for the wording of step 4b in Tactics Box 23.2 and is the third special ray shown in Figure 23.39.

You can see that the three refracted rays don't converge. Instead, all three rays appear to *diverge* from point P′. This is the situation we found for rays reflecting from

FIGURE 23.39 Rays from an object at distance $s < f$ are refracted by the lens and diverge to form a virtual image.

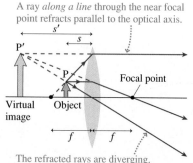

A ray *along a line* through the near focal point refracts parallel to the optical axis.

The refracted rays are diverging. They appear to come from point P′.

a mirror and for the rays refracting out of an aquarium. Point P′ is a *virtual image* of the object point P. Furthermore, it is an **upright image,** having the same orientation as the object.

The refracted rays, which are all to the right of the lens, *appear* to come from P′, but none of the rays were ever at that point. No image would appear on a screen placed in the image plane at P′. So what good is a virtual image?

Your eye collects and focuses bundles of diverging rays; thus, as **FIGURE 23.40a** shows, you can "see" a virtual image by looking *through* the lens. This is exactly what you do with a magnifying glass, producing a scene like the one in **FIGURE 23.40b**. In fact, you view a virtual image anytime you look *through* the eyepiece of an optical instrument such as a microscope or binoculars.

The image distance s′ for a virtual image is defined to be a *negative number* **($s′ < 0$)**, indicating that the image is on the opposite side of the lens from a real image. With this choice of sign, the definition of magnification, $m = -s′/s$, is still valid. A virtual image with negative $s′$ has $m > 0$, thus the image is upright. This agrees with the rays in Figure 23.39 and the photograph of Figure 23.40b.

NOTE ▶ A lens thicker in the middle than at the edges is classified as a converging lens. The light rays from an object *can* converge to form a real image after passing through such a lens, but only if the object distance is larger than the focal length of the lens: $s > f$. If $s < f$, the rays leaving a converging lens are diverging to produce a virtual image. ◀

FIGURE 23.40 A converging lens is a magnifying glass when the object distance is less than f.

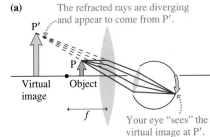
(a)

(b)

EXAMPLE 23.9 | **Magnifying a flower**

To see a flower better, a naturalist holds a 6.0-cm-focal-length magnifying glass 4.0 cm from the flower. What is the magnification?

MODEL The flower is in the object plane. Use ray tracing to locate the image.

VISUALIZE **FIGURE 23.41** shows the ray-tracing diagram. The three special rays diverge from the lens, but we can use a straightedge to extend the rays backward to the point from which they diverge. This point, the image point, is seen to be 12 cm to the left of the lens. Because this is a virtual image, the image distance is $s′ = -12$ cm. Thus the magnification is

$$m = -\frac{s′}{s} = -\frac{-12 \text{ cm}}{4.0 \text{ cm}} = 3.0$$

The image is three times as large as the object and, because m is positive, upright.

FIGURE 23.41 Ray-tracing diagram for Example 23.9.

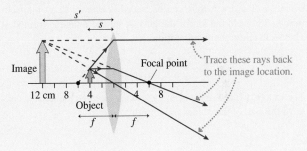

Diverging Lenses

A lens thicker at the edges than in the middle is called a *diverging lens.* **FIGURE 23.42** shows three important sets of rays passing through a diverging lens. These are based on Figures 23.33 and 23.34, where you saw that rays initially parallel to the axis diverge after passing through a diverging lens.

FIGURE 23.42 Three important sets of rays passing through a thin diverging lens.

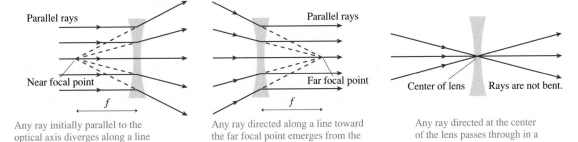

Any ray initially parallel to the optical axis diverges along a line through the near focal point.

Any ray directed along a line toward the far focal point emerges from the lens parallel to the optical axis.

Any ray directed at the center of the lens passes through in a straight line.

Ray tracing follows the steps of Tactics Box 23.2 for a converging lens *except* that two of the three special rays in step 4 are different.

TACTICS BOX 23.3 **Ray tracing for a diverging lens**

❶–❸ **Follow steps 1 through 3 of Tactics Box 23.2.**
❹ **Draw the three "special rays" from the tip of the arrow.** Use a straight-edge.

　　a. A ray parallel to the axis diverges along a line through the near focal point.
　　b. A ray along a line toward the far focal point emerges parallel to the axis.
　　c. A ray through the center of the lens does not bend.

❺ **Trace the diverging rays backward.** The point from which they are diverging is the image point, which is always a virtual image.
❻ **Measure the image distance s'.** This will be a negative number.

Exercise 28

EXAMPLE 23.10 **Demagnifying a flower**

A diverging lens with a focal length of 50 cm is placed 100 cm from a flower. Where is the image? What is its magnification?

MODEL The flower is in the object plane. Use ray tracing to locate the image.

VISUALIZE **FIGURE 23.43** shows the ray-tracing diagram. The three special rays (labeled a, b, and c to match the Tactics Box) do not converge. However, they can be traced backward to an intersection ≈ 33 cm to the left of the lens. A virtual image is formed at $s' = -33$ cm with magnification

$$m = -\frac{s'}{s} = -\frac{-33 \text{ cm}}{100 \text{ cm}} = 0.33$$

The image, which can be seen by looking *through* the lens, is one-third the size of the object and upright.

FIGURE 23.43 Ray-tracing diagram for Example 23.10.

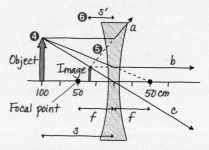

ASSESS Ray tracing with a diverging lens is somewhat trickier than with a converging lens, so this example is worth careful study.

Diverging lenses *always* make virtual images and, for this reason, are rarely used alone. However, they have important applications when used in combination with other lenses. Cameras, eyepieces, and eyeglasses often incorporate diverging lenses.

23.7 Thin Lenses: Refraction Theory

Ray tracing is a powerful visual approach for understanding image formation, but it doesn't provide precise information about the image location or image properties. We need to develop a quantitative relationship between the object distance s and the image distance s'.

To begin, **FIGURE 23.44** shows a *spherical* boundary between two transparent media with indices of refraction n_1 and n_2. The sphere has radius of curvature R. Consider a ray that leaves object point P at angle α and later, after refracting, reaches point P'. Figure 23.44 has exaggerated the angles to make the picture clear, but we will restrict our analysis to *paraxial rays* traveling nearly parallel to the axis. For paraxial rays, all the angles are small and we can use the small-angle approximation.

FIGURE 23.44 Image formation due to refraction at a spherical surface. The angles are exaggerated.

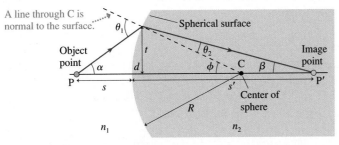

The ray from P is incident on the boundary at angle θ_1 and refracts into medium n_2 at angle θ_2, both measured from the normal to the surface at the point of incidence. Snell's law is $n_1 \sin\theta_1 = n_2 \sin\theta_2$, which in the small-angle approximation is

$$n_1\theta_1 = n_2\theta_2 \tag{23.15}$$

You can see from the geometry of Figure 23.44 that angles α, β, and ϕ are related by

$$\begin{aligned}\theta_1 &= \alpha + \phi \\ \theta_2 &= \phi - \beta\end{aligned} \tag{23.16}$$

Using these expressions in Equation 23.15, we can write Snell's law as

$$n_1(\alpha + \phi) = n_2(\phi - \beta) \tag{23.17}$$

This is one important relationship between the angles.

The line of height t, from the axis to the point of incidence, is the vertical leg of three different right triangles having vertices at points P, C, and P′. Consequently,

$$\tan\alpha \approx \alpha = \frac{t}{s+d} \qquad \tan\beta \approx \beta = \frac{t}{s'-d} \qquad \tan\phi \approx \phi = \frac{t}{R-d} \tag{23.18}$$

But $d \to 0$ for paraxial rays, thus

$$\alpha = \frac{t}{s} \qquad \beta = \frac{t}{s'} \qquad \phi = \frac{t}{R} \tag{23.19}$$

This is the second important relationship that comes from Figure 23.44.

If we use the angles of Equation 23.19 in Equation 23.17, we find

$$n_1\left(\frac{t}{s} + \frac{t}{R}\right) = n_2\left(\frac{t}{R} - \frac{t}{s'}\right) \tag{23.20}$$

The t cancels, and we can rearrange Equation 23.20 to read

$$\frac{n_1}{s} + \frac{n_2}{s'} = \frac{n_2 - n_1}{R} \tag{23.21}$$

Equation 23.21 is independent of angle α. Consequently, **all paraxial rays leaving point P later converge at point P′.** If an object is located at distance s from a spherical refracting surface, an image will be formed at distance s' given by Equation 23.21.

Equation 23.21 was derived for a surface that is convex toward the object point, and the image is real. However, the result is also valid for virtual images or for surfaces that are concave toward the object point as long as we adopt the *sign convention* shown in Table 23.3.

Section 23.4 considered image formation due to refraction by a plane surface. There we found (in Equation 23.13) an image distance $s' = (n_2/n_1)s$. A plane can be thought of as a sphere in the limit $R \to \infty$, so we should be able to reach the same conclusion from Equation 23.21. As $R \to \infty$, the term $(n_2 - n_1)/R \to 0$ and Equation 23.21 becomes $s' = -(n_2/n_1)s$. This seems to differ from Equation 23.13, but it

TABLE 23.3 Sign convention for refracting surfaces

	Positive	Negative
R	Convex toward the object	Concave toward the object
s'	Real image, opposite side from object	Virtual image, same side as object

doesn't really. Equation 23.13 gives the actual distance to the image. Equation 23.21 is based on a sign convention in which virtual images have negative image distances, hence the minus sign.

EXAMPLE 23.11 | **Image formation inside a glass rod**

One end of a 4.0-cm-diameter glass rod is shaped like a hemisphere. A small lightbulb is 6.0 cm from the end of the rod. Where is the bulb's image located?

MODEL Model the lightbulb as a point source of light and consider the paraxial rays that refract into the glass rod.

FIGURE 23.45 The curved surface refracts the light to form a real image.

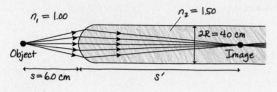

VISUALIZE FIGURE 23.45 shows the situation. $n_1 = 1.00$ for air and $n_2 = 1.50$ for glass.

SOLVE The radius of the surface is half the rod diameter, so $R = 2.0$ cm. Equation 23.21 is

$$\frac{1.00}{6.0 \text{ cm}} + \frac{1.50}{s'} = \frac{1.50 - 1.00}{2.0 \text{ cm}} = \frac{0.50}{2.0 \text{ cm}}$$

Solving for the image distance s' gives

$$\frac{1.50}{s'} = \frac{0.50}{2.0 \text{ cm}} - \frac{1.00}{6.0 \text{ cm}} = 0.0833 \text{ cm}^{-1}$$

$$s' = \frac{1.50}{0.0833} = 18 \text{ cm}$$

ASSESS This is a real image located 18 cm inside the glass rod.

EXAMPLE 23.12 | **A goldfish in a bowl**

A goldfish lives in a spherical fish bowl 50 cm in diameter. If the fish is 10 cm from the near edge of the bowl, where does the fish appear when viewed from the outside?

MODEL Model the fish as a point source and consider the paraxial rays that refract from the water into the air. The thin glass wall has little effect and will be ignored.

FIGURE 23.46 The curved surface of a fish bowl produces a virtual image of the fish.

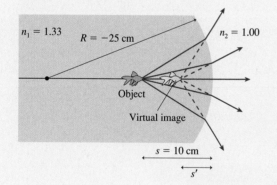

VISUALIZE FIGURE 23.46 shows the rays refracting *away* from the normal as they move from the water into the air. We expect to find a virtual image at a distance less than 10 cm.

SOLVE The object is in the water, so $n_1 = 1.33$ and $n_2 = 1.00$. The inner surface is concave (you can remember "concave" because it's like looking into a cave), so $R = -25$ cm. The object distance is $s = 10$ cm. Thus Equation 23.21 is

$$\frac{1.33}{10 \text{ cm}} + \frac{1.00}{s'} = \frac{1.00 - 1.33}{-25 \text{ cm}} = \frac{0.33}{25 \text{ cm}}$$

Solving for the image distance s' gives

$$\frac{1.00}{s'} = \frac{0.33}{25 \text{ cm}} - \frac{1.33}{10 \text{ cm}} = -0.12 \text{ cm}^{-1}$$

$$s' = \frac{1.00}{-0.12 \text{ cm}^{-1}} = -8.3 \text{ cm}$$

ASSESS The image is virtual, located to the left of the boundary. A person looking into the bowl will see a fish that appears to be 8.3 cm from the edge of the bowl.

STOP TO THINK 23.5 Which of these actions will move the real image point P′ farther from the boundary? More than one may work.

a. Increase the radius of curvature R.
b. Increase the index of refraction n.
c. Increase the object distance s.
d. Decrease the radius of curvature R.
e. Decrease the index of refraction n.
f. Decrease the object distance s.

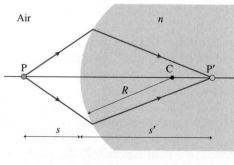

Lenses

The thin-lens approximation assumes rays refract one time, at the lens plane. In fact, as FIGURE 23.47 shows, rays refract *twice,* at spherical surfaces having radii of curvature R_1 and R_2. Let the lens have thickness t and be made of a material with index of refraction n. For simplicity, we'll assume that the lens is surrounded by air.

FIGURE 23.47 Image formation by a lens.

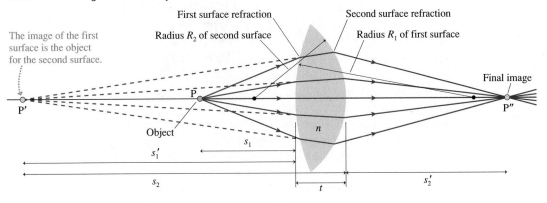

The object at point P is distance s_1 to the left of the lens. The first surface of the lens, of radius R_1, refracts the rays from P to create an image at point P′. We can use Equation 23.21 for a spherical surface to find the image distance s_1':

$$\frac{1}{s_1} + \frac{n}{s_1'} = \frac{n-1}{R_1} \tag{23.22}$$

where we used $n_1 = 1$ for the air and $n_2 = n$ for the lens. We'll assume that the image P′ is a virtual image, but this assumption isn't essential to the outcome.

With two refracting surfaces, the image P′ of the first surface becomes the object for the second surface. That is, the rays refracting at the second surface appear to have come from P′. Object distance s_2 from P′ to the second surface looks like it should be $s_2 = s_1' + t$, but P′ is a virtual image, so s_1' is a *negative* number. Thus the distance to the second surface is $s_2 = |s_1'| + t = t - s_1'$. We can find the image of P′ by a second application of Equation 23.21, but with a switch. The rays are incident on the surface from within the lens, so this time $n_1 = n$ and $n_2 = 1$. Consequently,

$$\frac{n}{t - s_1'} + \frac{1}{s_2'} = \frac{1-n}{R_2} \tag{23.23}$$

For a *thin lens,* which has $t \to 0$, Equation 23.23 becomes

$$-\frac{n}{s_1'} + \frac{1}{s_2'} = \frac{1-n}{R_2} = -\frac{n-1}{R_2} \tag{23.24}$$

Our goal is to find the distance s_2' to point P″, the image produced by the lens as a whole. This goal is easily reached if we simply add Equations 23.22 and 23.24, eliminating s_1' and giving

$$\frac{1}{s_1} + \frac{1}{s_2'} = \frac{n-1}{R_1} - \frac{n-1}{R_2} = (n-1)\left(\frac{1}{R_1} - \frac{1}{R_2}\right) \tag{23.25}$$

The numerical subscripts on s_1 and s_2' no longer serve a purpose. If we replace s_1 by s, the object distance from the lens, and s_2' by s', the image distance, Equation 23.25 becomes the *thin-lens equation:*

$$\frac{1}{s} + \frac{1}{s'} = \frac{1}{f} \qquad \text{(thin-lens equation)} \qquad (23.26)$$

where the *focal length* of the lens is

$$\frac{1}{f} = (n - 1)\left(\frac{1}{R_1} - \frac{1}{R_2}\right) \qquad \text{(lens maker's equation)} \qquad (23.27)$$

Equation 23.27 is known as the *lens maker's equation.* It allows you to determine the focal length from the shape of a thin lens and the material used to make it.

We can verify that this expression for f really is the focal length of the lens by recalling that rays initially parallel to the optical axis pass through the focal point on the far side. In fact, this was our *definition* of the focal length of a lens. Parallel rays must come from an object extremely far away, with object distance $s \rightarrow \infty$ and thus $1/s = 0$. In that case, Equation 23.26 tells us that the parallel rays will converge at distance $s' = f$ on the far side of the lens, exactly as expected.

We derived the thin-lens equation and the lens maker's equation from the specific lens geometry shown in Figure 23.47, but the results are valid for any lens as long as all quantities are given appropriate signs. The sign convention used with Equations 23.26 and 23.27 is given in Table 23.4.

TABLE 23.4 Sign convention for thin lenses

	Positive	Negative
R_1, R_2	Convex toward the object	Concave toward the object
f	Converging lens, thicker in center	Diverging lens, thinner in center
s'	Real image, opposite side from object	Virtual image, same side as object

NOTE ▶ For a *thick lens,* where the thickness t is not negligible, we can solve Equations 23.22 and 23.23 in sequence to find the position of the image point P''. ◀

EXAMPLE 23.13 **Focal length of a meniscus lens**

What is the focal length of the glass *meniscus lens* shown in **FIGURE 23.48**? Is this a converging or diverging lens?

FIGURE 23.48 A meniscus lens.

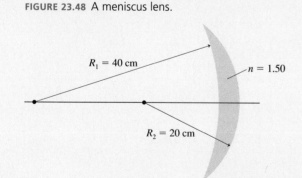

$R_1 = 40$ cm

$n = 1.50$

$R_2 = 20$ cm

SOLVE If the object is on the left, then the first surface has $R_1 = -40$ cm (concave toward the object) and the second surface has $R_2 = -20$ cm (also concave toward the object). The index of refraction of glass is $n = 1.50$, so the lens maker's equation is

$$\frac{1}{f} = (n - 1)\left(\frac{1}{R_1} - \frac{1}{R_2}\right) = (1.50 - 1)\left(\frac{1}{-40 \text{ cm}} - \frac{1}{-20 \text{ cm}}\right)$$
$$= 0.0125 \text{ cm}^{-1}$$

Inverting this expression gives $f = 80$ cm. This is a converging lens, as seen both from the positive value of f and from the fact that the lens is thicker in the center.

Thin-Lens Image Formation

Although the thin-lens equation allows precise calculations, the lessons of ray tracing should not be forgotten. The most powerful tool of optical analysis is a combination of ray tracing, to gain an intuitive understanding of the ray trajectories, and the thin-lens equation.

EXAMPLE 23.14 **Designing a lens**

The objective lens of a microscope uses a planoconvex glass lens with the flat side facing the specimen. A real image is formed 160 mm behind the lens when the lens is 8.0 mm from the specimen. What is the radius of the lens's curved surface?

MODEL Treat the lens as a thin lens with the specimen as the object. The lens's focal length is given by the lens maker's equation.

VISUALIZE **FIGURE 23.49** clarifies the shape of the lens and defines R_2. The index of refraction was taken from Table 23.1.

FIGURE 23.49 A planoconvex microscope lens.

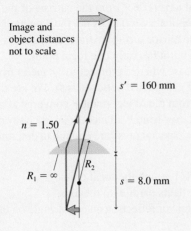

Image and object distances not to scale

$s' = 160$ mm

$n = 1.50$

$R_1 = \infty$

R_2

$s = 8.0$ mm

SOLVE We can use the lens maker's equation to solve for R_2 if we know the lens's focal length. Because we know both the object and image distances, we can use the thin-lens equation to find

$$\frac{1}{f} = \frac{1}{s} + \frac{1}{s'} = \frac{1}{8.0 \text{ mm}} + \frac{1}{160 \text{ mm}} = 0.131 \text{ mm}^{-1}$$

The focal length is $f = 1/(0.131 \text{ mm}^{-1}) = 7.6$ mm, but $1/f$ is all we need for the lens maker's equation. The front surface of the lens is planar, which we can consider a portion of a sphere with $R_1 \rightarrow \infty$. Consequently $1/R_1 = 0$. With this, we can solve the lens maker's equation for R_2:

$$\frac{1}{R_2} = \frac{1}{R_1} - \frac{1}{n-1}\frac{1}{f} = 0 - \left(\frac{1}{1.50 - 1}\right)(0.131 \text{ mm}^{-1})$$

$$= -0.262 \text{ mm}^{-1}$$

$$R_2 = -3.8 \text{ mm}$$

The minus sign appears because the curved surface is concave toward the object. Physically, the radius of the curved surface is 3.8 mm.

ASSESS The actual thickness of the lens has to be less than R_2, probably no more than about 1.0 mm. This thickness is significantly less than the object and image distances, so the thin-lens approximation is justified.

EXAMPLE 23.15 **A magnifying lens**

A stamp collector uses a magnifying lens that sits 2.0 cm above the stamp. The magnification is 4.0. What is the focal length of the lens?

FIGURE 23.50 Pictorial representation of a magnifying lens.

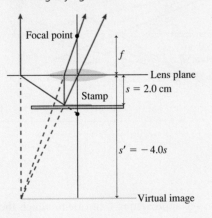

Focal point

f

Lens plane

$s = 2.0$ cm

Stamp

$s' = -4.0s$

Virtual image

MODEL A magnifying lens is a converging lens with the object distance less than the focal length ($s < f$). Assume it is a thin lens.

VISUALIZE **FIGURE 23.50** shows the lens and a ray-tracing diagram. We do not need to know the actual shape of the lens, so the figure shows a generic converging lens.

SOLVE A virtual image is upright, so $m = +4.0$. The magnification is $m = -s'/s$, thus

$$s' = -4.0s = -(4.0)(2.0 \text{ cm}) = -8.0 \text{ cm}$$

We can use s and s' in the thin-lens equation to find the focal length:

$$\frac{1}{f} = \frac{1}{s} + \frac{1}{s'} = \frac{1}{2.0 \text{ cm}} + \frac{1}{-8.0 \text{ cm}} = 0.375 \text{ cm}^{-1}$$

$$f = 2.7 \text{ cm}$$

ASSESS $f > 2$ cm, as expected.

STOP TO THINK 23.6 A lens forms a real image of a lightbulb, but the image of the bulb on a viewing screen is blurry because the screen is slightly in front of the image plane. To focus the image, should you move the lens toward the bulb or away from the bulb?

23.8 Image Formation with Spherical Mirrors

Curved mirrors—such as those used in telescopes, security and rearview mirrors, and searchlights—can be used to form images, and their images can be analyzed with ray diagrams similar to those used with lenses. We'll consider only the important case of **spherical mirrors,** whose surface is a section of a sphere.

Concave Mirrors

FIGURE 23.51 shows a **concave mirror,** a mirror in which the edges curve *toward* the light source. Rays parallel to the optical axis reflect from the surface of the mirror so as to pass through a single point on the optical axis. This is the focal point of the mirror. The focal length is the distance from the mirror surface to the focal point. A concave mirror is analogous to a converging lens, but it has only one focal point.

Let's begin by considering the case where the object's distance s from the mirror is greater than the focal length ($s > f$), as shown in **FIGURE 23.52**. We see that the image is *real* (and inverted) because rays from the object point P converge at the image point P′. Although an infinite number of rays from P all meet at P′, each ray obeying the law of reflection, you can see that three "special rays" are enough to determine the position and size of the image:

- A ray parallel to the axis reflects through the focal point.
- A ray through the focal point reflects parallel to the axis.
- A ray striking the center of the mirror reflects at an equal angle on the opposite side of the axis.

These three rays also locate the image if $s < f$, but in that case the image is *virtual* and behind the mirror.

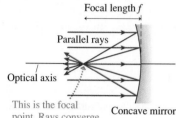

FIGURE 23.51 The focal point and focal length of a concave mirror.

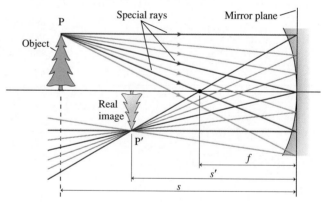

FIGURE 23.52 A real image formed by a concave mirror.

Convex Mirrors

FIGURE 23.53 shows parallel light rays approaching a mirror in which the edges curve *away from* the light source. This is called a **convex mirror.** In this case, the reflected rays appear to come from a point behind the mirror. This is the focal point for a convex mirror.

A common example of a convex mirror is a silvered ball, such as a tree ornament. You may have noticed that if you look at your reflection in such a ball, your image appears right-side-up but is quite small. As another example, **FIGURE 23.54** shows a city skyline reflected in a polished metal sphere. Let's use ray tracing to understand why the skyscrapers all appear to be so small.

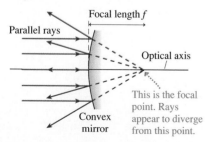

FIGURE 23.53 The focal point and focal length of a convex mirror.

FIGURE 23.55 shows an object in front of a convex mirror. In this case, the reflected rays—each obeying the law of reflection—create an upright image of reduced height behind the mirror. We see that the image is virtual because no rays actually converge at the image point P′. Instead, diverging rays *appear* to come from this point. Once again, three special rays are enough to find the image.

FIGURE 23.54 A city skyline is reflected in this polished sphere.

FIGURE 23.55 A virtual image formed by a convex mirror.

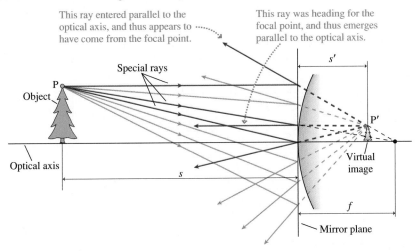

Convex mirrors are used for a variety of safety and monitoring applications, such as passenger-side rearview mirrors and the round mirrors used in stores to keep an eye on the customers. When an object is reflected in a convex mirror, the image appears smaller than the object itself. Because the image is, in a sense, a miniature version of the object, you can *see much more of it* within the edges of the mirror than you could with an equal-sized flat mirror.

TACTICS
BOX 23.4 **Ray tracing for a spherical mirror**

❶ **Draw an optical axis.** Use graph paper or a ruler! Establish an appropriate scale.

❷ **Center the mirror on the axis.** Mark and label the focal point at distance f from the mirror's surface.

❸ **Represent the object with an upright arrow at distance s.** It's usually best to place the base of the arrow on the axis and to draw the arrow about half the radius of the mirror.

❹ **Draw the three "special rays" from the tip of the arrow.** Use a straight-edge.

    a. A ray parallel to the axis reflects through (concave) or away from (convex) the focal point.

    b. An incoming ray passing through (concave) or heading toward (convex) the focal point reflects parallel to the axis.

    c. A ray that strikes the center of the mirror reflects at an equal angle on the opposite side of the optical axis.

❺ **Extend the rays forward or backward until they converge.** This is the image point. Draw the rest of the image in the image plane. If the base of the object is on the axis, then the base of the image will also be on the axis.

❻ **Measure the image distance s'.** Also, if needed, measure the image height relative to the object height.

Exercises 32–33

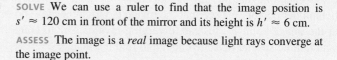

EXAMPLE 23.16 | Analyzing a concave mirror

A 3.0-cm-high object is located 60 cm from a concave mirror. The mirror's focal length is 40 cm. Use ray tracing to find the position and height of the image.

MODEL Use the ray-tracing steps of Tactics Box 23.4.

VISUALIZE **FIGURE 23.56** shows the steps of Tactics Box 23.4.

SOLVE We can use a ruler to find that the image position is $s' \approx 120$ cm in front of the mirror and its height is $h' \approx 6$ cm.

ASSESS The image is a *real* image because light rays converge at the image point.

FIGURE 23.56 Ray-tracing diagram for a concave mirror.

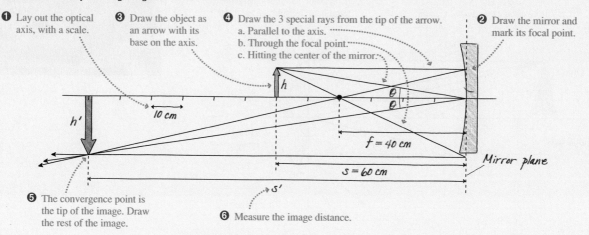

❶ Lay out the optical axis, with a scale.
❸ Draw the object as an arrow with its base on the axis.
❹ Draw the 3 special rays from the tip of the arrow.
  a. Parallel to the axis.
  b. Through the focal point.
  c. Hitting the center of the mirror.
❷ Draw the mirror and mark its focal point.
❺ The convergence point is the tip of the image. Draw the rest of the image.
❻ Measure the image distance.

The Mirror Equation

The thin-lens equation assumes lenses have negligible thickness (so a single refraction occurs in the lens plane) and the rays are nearly parallel to the optical axis (paraxial rays). If we make the same assumptions about spherical mirrors—the mirror has negligible thickness and so paraxial rays reflect at the mirror plane—then the object and image distances are related exactly as they were for thin lenses:

$$\frac{1}{s} + \frac{1}{s'} = \frac{1}{f} \qquad \text{(mirror equation)} \qquad (23.28)$$

The focal length of the mirror, as you can show as a homework problem, is related to the mirror's radius of curvature by

$$f = \frac{R}{2} \qquad (23.29)$$

TABLE 23.5 Sign convention for spherical mirrors

	Positive	Negative
R, f	Concave toward the object	Convex toward the object
s'	Real image, same side as object	Virtual image, opposite side from object

Table 23.5 shows the sign convention used with spherical mirrors. It differs from the convention for lenses, so you'll want to carefully compare this table to Table 23.4. A concave mirror (analogous to a converging lens) has a positive focal length while a convex mirror (analogous to a diverging lens) has a negative focal length. The lateral magnification of a spherical mirror is computed exactly as for a lens:

$$m = -\frac{s'}{s} \qquad (23.30)$$

EXAMPLE 23.17 | Analyzing a concave mirror

A 3.0-cm-high object is located 20 cm from a concave mirror. The mirror's radius of curvature is 80 cm. Determine the position, orientation, and height of the image.

MODEL Treat the mirror as a thin mirror.

VISUALIZE The mirror's focal length is $f = R/2 = +40$ cm, where we used the sign convention from Table 23.5. With the focal length known, the three special rays in **FIGURE 23.57** show that the image is a magnified, virtual image behind the mirror.

FIGURE 23.57 Pictorial representation of Example 23.17.

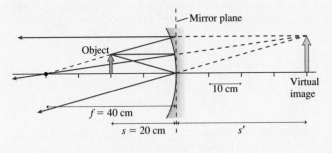

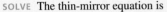

SOLVE The thin-mirror equation is

$$\frac{1}{20 \text{ cm}} + \frac{1}{s'} = \frac{1}{40 \text{ cm}}$$

This is easily solved to give $s' = -40$ cm, in agreement with the ray tracing. The negative sign tells us this is a virtual image behind the mirror. The magnification is

$$m = -\frac{-40 \text{ cm}}{20 \text{ cm}} = +2.0$$

Consequently, the image is 6.0 cm tall and upright.

ASSESS This is a virtual image because light rays diverge from the image point. You could see this enlarged image by standing behind the object and looking into the mirror. In fact, this is how magnifying cosmetic mirrors work.

STOP TO THINK 23.7 A concave mirror of focal length f forms an image of the moon. Where is the image located?

a. At the mirror's surface
b. Almost exactly a distance f behind the mirror
c. Almost exactly a distance f in front of the mirror
d. At a distance behind the mirror equal to the distance of the moon in front of the mirror

---

CHALLENGE EXAMPLE 23.18 **Optical fiber imaging**

An *endoscope* is a thin bundle of optical fibers that can be inserted through a bodily opening or small incision to view the interior of the body. As FIGURE 23.58 shows, an *objective* lens forms a real image on the entrance face of the fiber bundle. Individual fibers, using total internal reflection, transport the light to the exit face, where it emerges. The doctor (or a TV camera) observes the object by viewing the exit face through an *eyepiece* lens.

Consider an endoscope having a 3.0-mm-diameter objective lens with a focal length of 1.1 mm. These are typical values. The indices of refraction of the core and the cladding of the optical fibers are 1.62 and 1.50, respectively. To give maximum brightness, the objective lens is positioned so that, for an on-axis object, rays passing through the outer edge of the lens have the maximum angle of incidence for undergoing TIR in the fiber. How far should the objective lens be placed from the object the doctor wishes to view?

MODEL Represent the object as an on-axis point source and use the ray model of light.

VISUALIZE FIGURE 23.59 on the next page shows the real image being focused on the entrance face of the endoscope. Inside the fiber, rays that strike the cladding at an angle of incidence greater than the critical angle θ_c undergo TIR and stay in the fiber; rays are lost if their angle of incidence is less than θ_c. For maximum brightness, the lens is positioned so that a ray passing through the outer edge refracts into the fiber at the maximum angle of incidence θ_{max} for which TIR is possible. A smaller-diameter lens would sacrifice light-gathering power, whereas the outer rays from a larger-diameter lens would impinge on the core-cladding boundary at less than θ_c and would not undergo TIR. Note that the lens-to-fiber distance, although unknown, is fixed by the manufacturer and cannot be changed. Only object distance is under the doctor's control.

FIGURE 23.58 An endoscope.

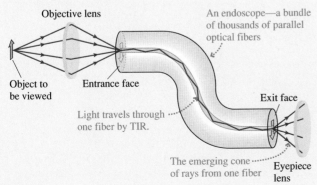

Continued

FIGURE 23.59 Magnified view of the entrance of an optical fiber.

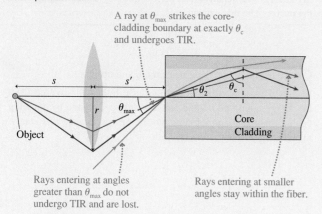

A ray at θ_{max} strikes the core-cladding boundary at exactly θ_c and undergoes TIR.

Object

Rays entering at angles greater than θ_{max} do not undergo TIR and are lost.

Rays entering at smaller angles stay within the fiber.

Core
Cladding

SOLVE We know the focal length of the lens. We can use the geometry of the ray at the critical angle to find the image distance s', then use the thin-lens equation to find the object distance s. The critical angle for TIR inside the fiber is

$$\theta_c = \sin^{-1}\left(\frac{n_{cladding}}{n_{core}}\right) = \sin^{-1}\left(\frac{1.50}{1.62}\right) = 67.8°$$

A ray incident on the core-cladding boundary at exactly the critical angle must have entered the fiber, at the entrance face, at angle

$\theta_2 = 90° - \theta_c = 22.2°$. For optimum lens placement, this ray passed through the outer edge of the lens and was incident on the entrance face at angle θ_{max}. Snell's law at the entrance face is

$$n_{air} \sin \theta_{max} = 1.0 \cdot \sin \theta_{max} = n_{core} \sin \theta_2$$

and thus

$$\theta_{max} = \sin^{-1}(1.62 \sin 22.2°) = 37.7°$$

We know the lens radius, $r = 1.5$ mm, so the distance of the lens from the fiber—the image distance s'—is

$$s' = \frac{r}{\tan \theta_{max}} = \frac{1.5 \text{ mm}}{\tan(37.7°)} = 1.9 \text{ mm}$$

Now we can use the thin-lens equation to locate the object:

$$\frac{1}{s} = \frac{1}{f} - \frac{1}{s'} = \frac{1}{1.1 \text{ mm}} - \frac{1}{1.9 \text{ mm}}$$

$$s = 2.6 \text{ mm}$$

The doctor, viewing the exit face of the fiber bundle, will see a focused image when the objective lens is 2.6 mm from the object she wishes to view.

ASSESS The object and image distances are both greater than the focal length, which is correct for forming a real image.

SUMMARY

The goals of Chapter 23 have been to understand and apply the ray model of light.

General Principles

Reflection

Law of reflection: $\theta_r = \theta_i$

Reflection can be **specular** (mirror-like) or **diffuse** (from rough surfaces).

Plane mirrors: A virtual image is formed at P' with $s' = s$.

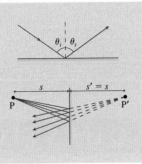

Refraction

Snell's law of refraction:

$$n_1 \sin\theta_1 = n_2 \sin\theta_2$$

Index of refraction is $n = c/v$. The ray is closer to the normal on the side with the larger index of refraction.

If $n_2 < n_1$, **total internal reflection** (TIR) occurs when the angle of incidence $\theta_1 \geq \theta_c = \sin^{-1}(n_2/n_1)$.

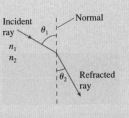

Important Concepts

The ray model of light

Light travels along straight lines, called **light rays,** at speed $v = c/n$.

A light ray continues forever unless an interaction with matter causes it to reflect, refract, scatter, or be absorbed.

Light rays come from **objects.** Each point on the object sends rays in all directions.

The eye sees an object (or an image) when diverging rays are collected by the pupil and focused on the retina.

▶ Ray optics is valid when lenses, mirrors, and apertures are larger than ≈ 1 mm.

Image formation

If rays diverge from P and interact with a lens or mirror so that the refracted rays *converge* at P', then P' is a real image of P.

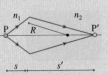

If rays diverge from P and interact with a lens or mirror so that the refracted/reflected rays *diverge* from P' and appear to come from P', then P' is a virtual image of P.

Spherical surface: Object and image distances are related by

$$\frac{n_1}{s} + \frac{n_2}{s'} = \frac{n_2 - n_1}{R}$$

Plane surface: $R \to \infty$, so $|s'/s| = n_2/n_1$.

Applications

Ray tracing

3 special rays in 3 basic situations:

Converging lens   Converging lens   Diverging lens
Real image   Virtual image   Virtual image

Magnification $m = -\dfrac{s'}{s}$

m is $+$ for an upright image, $-$ for inverted.
The height ratio is $h'/h = |m|$.

Thin lenses

The image and object distances are related by

$$\frac{1}{s} + \frac{1}{s'} = \frac{1}{f}$$

Focal length f

where the focal length is given by the lens maker's equation:

$$\frac{1}{f} = (n - 1)\left(\frac{1}{R_1} - \frac{1}{R_2}\right)$$

R $+$ for surface convex toward object	$-$ for concave
f $+$ for a converging lens	$-$ for diverging
s' $+$ for a real image	$-$ for virtual

Spherical mirrors

The image and object distances are related by

$$\frac{1}{s} + \frac{1}{s'} = \frac{1}{f}$$

R, f $+$ for concave mirror	$-$ for convex
s' $+$ for a real image	$-$ for virtual

Focal length $f = R/2$

Terms and Notation

light ray	diffuse reflection	dispersion	object plane
object	virtual image	Rayleigh scattering	image plane
point source	refraction	lens	inverted image
parallel bundle	angle of refraction	ray tracing	lateral magnification, m
ray diagram	Snell's law	converging lens	upright image
camera obscura	total internal reflection (TIR)	focal point	spherical mirror
aperture	critical angle, θ_c	focal length, f	concave mirror
specular reflection	object distance, s	diverging lens	convex mirror
angle of incidence	image distance, s'	thin lens	
angle of reflection	optical axis	lens plane	
law of reflection	paraxial rays	real image	

CONCEPTUAL QUESTIONS

1. If you turn on your car headlights during the day, the road ahead of you doesn't appear to get brighter. Why not?

2. Suppose you have two pinhole cameras. The first has a small round hole in the front. The second is identical except it has a square hole of the same area as the round hole in the first camera. Would the pictures taken by these two cameras, under the same conditions, be different in any obvious way? Explain.

3. You are looking at the image of a pencil in a mirror, as shown in **FIGURE Q23.3**.
   a. What happens to the image if the top half of the mirror, down to the midpoint, is covered with a piece of cardboard? Explain.
   b. What happens to the image if the bottom half of the mirror is covered with a piece of cardboard? Explain.

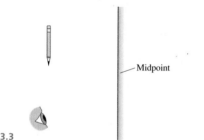

FIGURE Q23.3

4. One problem with using optical fibers for communication is that a light ray passing directly down the center of the fiber takes less time to travel from one end to the other than a ray taking a longer, zig-zag path. Thus light rays starting at the same time but traveling in slightly different directions reach the end of the fiber at different times. This problem can be solved by making the refractive index of the glass change gradually from a higher value in the center to a lower value near the edges of the fiber. Explain how this reduces the difference in travel times.

5. Suppose you looked at the sky on a clear day through pieces of red and blue plastic oriented as shown in **FIGURE Q23.5**. Describe the color and brightness of the light coming through sections 1, 2, and 3.

FIGURE Q23.5

6. A red card is illuminated by red light. What color will the card appear? What if it's illuminated by blue light?

7. The center of the galaxy is filled with low-density hydrogen gas. An astronomer wants to take a picture of the center of the galaxy. Will the view be better using ultraviolet light, visible light, or infrared light? (High-quality telescopes are available in all three spectral regions.) Explain.

8. Consider *one* point on an object near a lens.
   a. What is the minimum number of rays needed to locate its image point? Explain.
   b. How many rays from this point actually strike the lens and refract to the image point?

9. The object and lens in **FIGURE Q23.9** are positioned to form a well-focused, inverted image on a viewing screen. Then a piece of cardboard is lowered just in front of the lens to cover the top half of the lens. Describe what you see on the screen when the cardboard is in place.

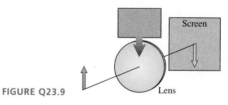

FIGURE Q23.9

10. **FIGURE Q23.10** shows an object near a lens. The focal points are marked. Is there an image? If so, is the image real or virtual? Is it upright or inverted? If not, why not? Explain.

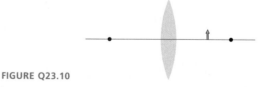

FIGURE Q23.10

11. A concave mirror brings the sun's rays to a focus in front of the mirror. Suppose the mirror is submerged in a swimming pool but still pointed up at the sun. Will the sun's rays be focused nearer to, farther from, or at the same distance from the mirror? Explain.

12. When you look at your reflection in the bowl of a spoon, it is upside down. Why?

EXERCISES AND PROBLEMS

Exercises

Section 23.1 The Ray Model of Light

1. ‖ a. How long (in ns) does it take light to travel 1.0 m in vacuum?
   b. What distance does light travel in water, glass, and cubic zirconia during the time that it travels 1.0 m in vacuum?

2. ‖ A point source of light illuminates an aperture 2.0 m away. A 12.0-cm-wide bright patch of light appears on a screen 1.0 m behind the aperture. How wide is the aperture?

3. ‖ A 5.0-cm-thick layer of oil is sandwiched between a 1.0-cm-thick sheet of glass and a 2.0-cm-thick sheet of polystyrene plastic. How long (in ns) does it take light incident perpendicular to the glass to pass through this 8.0-cm-thick sandwich?

4. ‖ A student has built a 15-cm-long pinhole camera for a science fair project. She wants to photograph her 180-cm-tall friend and have the image on the film be 5.0 cm high. How far should the front of the camera be from her friend?

Section 23.2 Reflection

5. | The mirror in FIGURE EX23.5 deflects a horizontal laser beam by 60°. What is the angle ϕ?

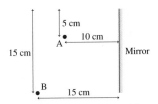

FIGURE EX23.5

6. | A light ray leaves point A in FIGURE EX23.6, reflects from the mirror, and reaches point B. How far below the top edge does the ray strike the mirror?

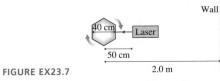

FIGURE EX23.6

7. ‖ The laser beam in FIGURE EX23.7 is aimed at the center of a rotating hexagonal mirror. How long is the streak of laser light as the reflected laser beam sweeps across the wall behind the laser?

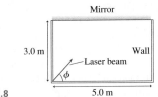

FIGURE EX23.7

8. ‖ At what angle ϕ should the laser beam in FIGURE EX23.8 be aimed at the mirrored ceiling in order to hit the midpoint of the far wall?

FIGURE EX23.8

9. ‖ It is 165 cm from your eyes to your toes. You're standing 200 cm in front of a tall mirror. How far is it from your eyes to the image of your toes?

Section 23.3 Refraction

10. ‖ A 1.0-cm-thick layer of water stands on a horizontal slab of glass. A light ray in the air is incident on the water 60° from the normal. What is the ray's direction of travel in the glass?

11. ‖ A costume jewelry pendant made of cubic zirconia is submerged in oil. A light ray strikes one face of the zirconia crystal at an angle of incidence of 25°. Once inside, what is the ray's angle with respect to the face of the crystal?

12. ‖ An underwater diver sees the sun 50° above horizontal. How high is the sun above the horizon to a fisherman in a boat above the diver?

13. | A laser beam in air is incident on a liquid at an angle of 53° with respect to the normal. The laser beam's angle in the liquid is 35°. What is the liquid's index of refraction?

14. ‖ The glass core of an optical fiber has an index of refraction 1.60. The index of refraction of the cladding is 1.48. What is the maximum angle a light ray can make with the wall of the core if it is to remain inside the fiber?

15. ‖ A thin glass rod is submerged in oil. What is the critical angle for light traveling inside the rod?

Section 23.4 Image Formation by Refraction

16. ‖ A fish in a flat-sided aquarium sees a can of fish food on the counter. To the fish's eye, the can looks to be 30 cm outside the aquarium. What is the actual distance between the can and the aquarium? (You can ignore the thin glass wall of the aquarium.)

17. | A biologist keeps a specimen of his favorite beetle embedded in a cube of polystyrene plastic. The hapless bug appears to be 2.0 cm within the plastic. What is the beetle's actual distance beneath the surface?

18. | A 150-cm-tall diver is standing completely submerged on the bottom of a swimming pool full of water. You are sitting on the end of the diving board, almost directly over her. How tall does the diver appear to be?

19. ‖ To a fish in an aquarium, the 4.00-mm-thick walls appear to be only 3.50 mm thick. What is the index of refraction of the walls?

Section 23.5 Color and Dispersion

20. ‖ A sheet of glass has $n_{red} = 1.52$ and $n_{violet} = 1.55$. A narrow beam of white light is incident on the glass at 30°. What is the angular spread of the light inside the glass?

21. | A narrow beam of white light is incident on a sheet of quartz. The beam disperses in the quartz, with red light ($\lambda \approx 700$ nm) traveling at an angle of 26.3° with respect to the normal and violet light ($\lambda \approx 400$ nm) traveling at 25.7°. The index of refraction of quartz for red light is 1.45. What is the index of refraction of quartz for violet light?

22. ‖ A hydrogen discharge lamp emits light with two prominent wavelengths: 656 nm (red) and 486 nm (blue). The light enters a flint-glass prism perpendicular to one face and then refracts through the hypotenuse back into the air. The angle between these two faces is 35°.
    a. Use Figure 23.28 to estimate to ±0.002 the index of refraction of flint glass at these two wavelengths.
    b. What is the angle (in degrees) between the red and blue light as it leaves the prism?

23. ‖ Infrared telescopes, which use special infrared detectors, are able to peer farther into star-forming regions of the galaxy because infrared light is not scattered as strongly as is visible light by the tenuous clouds of hydrogen gas from which new stars are created. For what wavelength of light is the scattering only 1% that of light with a visible wavelength of 500 nm?

Section 23.6 Thin Lenses: Ray Tracing

24. ‖ An object is 20 cm in front of a converging lens with a focal length of 10 cm. Use ray tracing to determine the location of the image. Is the image upright or inverted?

25. ‖ An object is 30 cm in front of a converging lens with a focal length of 5 cm. Use ray tracing to determine the location of the image. Is the image upright or inverted?

26. ‖ An object is 6 cm in front of a converging lens with a focal length of 10 cm. Use ray tracing to determine the location of the image. Is the image upright or inverted?

27. ‖ An object is 15 cm in front of a diverging lens with a focal length of −15 cm. Use ray tracing to determine the location of the image. Is the image upright or inverted?

Section 23.7 Thin Lenses: Refraction Theory

28. ‖ Find the focal length of the glass lens in FIGURE EX23.28.

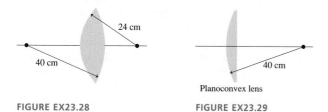

FIGURE EX23.28   FIGURE EX23.29

29. ‖ Find the focal length of the planoconvex polystyrene plastic lens in FIGURE EX23.29.
30. ‖ Find the focal length of the glass lens in FIGURE EX23.30.

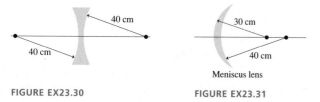

FIGURE EX23.30   FIGURE EX23.31

31. ‖ Find the focal length of the meniscus polystyrene plastic lens in FIGURE EX23.31.
32. ‖ An air bubble inside an 8.0-cm-diameter plastic ball is 2.0 cm from the surface. As you look at the ball with the bubble turned toward you, how far beneath the surface does the bubble appear to be?

33. ‖ A goldfish lives in a 50-cm-diameter spherical fish bowl. The fish sees a cat watching it. If the cat's face is 20 cm from the edge of the bowl, how far from the edge does the fish see it as being? (You can ignore the thin glass wall of the bowl.)
34. ‖ A 1.0-cm-tall candle flame is 60 cm from a lens with a focal length of 20 cm. What are the image distance and the height of the flame's image?

Section 23.8 Image Formation with Spherical Mirrors

35. ‖ An object is 40 cm in front of a concave mirror with a focal length of 20 cm. Use ray tracing to locate the image. Is the image upright or inverted?
36. ‖ An object is 12 cm in front of a concave mirror with a focal length of 20 cm. Use ray tracing to locate the image. Is the image upright or inverted?
37. ‖ An object is 30 cm in front of a convex mirror with a focal length of −20 cm. Use ray tracing to locate the image. Is the image upright or inverted?

Problems

38. ‖ An advanced computer sends information to its various parts via infrared light pulses traveling through silicon fibers. To acquire data from memory, the central processing unit sends a light-pulse request to the memory unit. The memory unit processes the request, then sends a data pulse back to the central processing unit. The memory unit takes 0.5 ns to process a request. If the information has to be obtained from memory in 2.0 ns, what is the maximum distance the memory unit can be from the central processing unit?

39. ‖ A red ball is placed at point A in FIGURE P23.39.
    a. How many images are seen by an observer at point O?
    b. What are the (x, y) coordinates of each image?

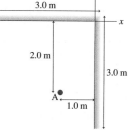

FIGURE P23.39

40. ‖ A laser beam is incident on the left mirror in FIGURE P23.40. Its initial direction is parallel to a line that bisects the mirrors. What is the angle φ of the reflected laser beam?

FIGURE P23.40

41. ‖ The place you get your hair cut has two nearly parallel mirrors 5.0 m apart. As you sit in the chair, your head is 2.0 m from the nearer mirror. Looking toward this mirror, you first see your face and then, farther away, the back of your head. (The mirrors need to be slightly nonparallel for you to be able to see the back of your head, but you can treat them as parallel in this problem.) How far away does the back of your head appear to be? Neglect the thickness of your head.

42. ‖ You're helping with an experiment in which a vertical cylinder will rotate about its axis by a very small angle. You need to devise a way to measure this angle. You decide to use what is called an *optical lever*. You begin by mounting a small mirror

on top of the cylinder. A laser 5.0 m away shoots a laser beam at the mirror. Before the experiment starts, the mirror is adjusted to reflect the laser beam directly back to the laser. Later, you measure that the reflected laser beam, when it returns to the laser, has been deflected sideways by 2.0 mm. Through how many degrees has the cylinder rotated?

43. ‖ A microscope is focused on a black dot. When a 1.00-cm-thick piece of plastic is placed over the dot, the microscope objective has to be raised 0.40 cm to bring the dot back into focus. What is the index of refraction of the plastic?

44. ‖ A light ray in air is incident on a transparent material whose index of refraction is n.
    a. Find an expression for the (non-zero) angle of incidence whose angle of refraction is half the angle of incidence.
    b. Evaluate your expression for light incident on glass.

45. ‖ The meter stick in **FIGURE P23.45** lies on the bottom of a 100-cm-long tank with its zero mark against the left edge. You look into the tank at a 30° angle, with your line of sight just grazing the upper left edge of the tank. What mark do you see on the meter stick if the tank is (a) empty, (b) half full of water, and (c) completely full of water?

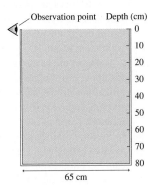

FIGURE P23.45

46. ‖ The 80-cm-tall, 65-cm-wide tank shown in **FIGURE P23.46** is completely filled with water. The tank has marks every 10 cm along one wall, and the 0 cm mark is barely submerged. As you stand beside the opposite wall, your eye is level with the top of the water.
    a. Can you see the marks from the top of the tank (the 0 cm mark) going down, or from the bottom of the tank (the 80 cm mark) coming up? Explain.
    b. Which is the lowest or highest mark, depending on your answer to part a, that you can see?

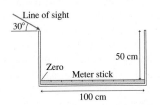

FIGURE P23.46

47. ‖ A 4.0-m-wide swimming pool is filled to the top. The bottom of the pool becomes completely shaded in the afternoon when the sun is 20° above the horizon. How deep is the pool?

48. ‖ It's nighttime, and you've dropped your goggles into a 3.0-m-deep swimming pool. If you hold a laser pointer 1.0 m above the edge of the pool, you can illuminate the goggles if the laser beam enters the water 2.0 m from the edge. How far are the goggles from the edge of the pool?

49. ‖ Shown from above in **FIGURE P23.49** is one corner of a rectangular box filled with water. A laser beam starts 10 cm from side A of the container and enters the water at position x. You can ignore the thin walls of the container.
    a. If $x = 15$ cm, does the laser beam refract back into the air through side B or reflect from side B back into the water? Determine the angle of refraction or reflection.

b. Repeat part a for $x = 25$ cm.
c. Find the minimum value of x for which the laser beam passes through side B and emerges into the air.

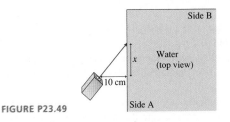

FIGURE P23.49

50. ‖ A fish is 20 m from the shore of a lake. A bonfire is burning on the edge of the lake nearest the fish.
    a. Does the fish need to be shallow (just below the surface) or very deep to see the light from the bonfire? Explain.
    b. What is the deepest or shallowest, depending on your answer to part a, that the fish can be and still see light from the fire?

51. ‖ Your supervisor asks you to measure the index of refraction of a piece of plastic. You notice that, because of scattering of the light, you can see the path of a laser beam through the plastic. You decide to shoot a laser beam toward the plastic at several different incident angles and measure the refraction angle in the plastic. Your data are as follows:

Incident angle	Refraction angle
15°	9°
30°	19°
45°	26°
60°	34°
75°	37°

Use the best-fit line of an appropriate graph to determine the plastic's index of refraction.

52. ‖‖ One of the contests at the school carnival is to throw a spear at an underwater target lying flat on the bottom of a pool. The water is 1.0 m deep. You're standing on a small stool that places your eyes 3.0 m above the bottom of the pool. As you look at the target, your gaze is 30° below horizontal. At what angle below horizontal should you throw the spear in order to hit the target? Your raised arm brings the spear point to the level of your eyes as you throw it, and over this short distance you can assume that the spear travels in a straight line rather than a parabolic trajectory.

53. ‖ White light is incident onto a 30° prism at the 40° angle shown in **FIGURE P23.53**. Violet light emerges perpendicular to the rear face of the prism. The index of refraction of violet light in this glass is 2.0% larger than the index of refraction of red light. At what angle ϕ does red light emerge from the rear face?

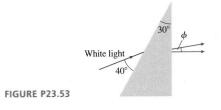

FIGURE P23.53

54. ‖ There's one angle of incidence β onto a prism for which the light inside an isosceles prism travels parallel to the base and emerges at angle β.

FIGURE P23.54

a. Find an expression for β in terms of the prism's apex angle α and index of refraction n.

b. A laboratory measurement finds that $\beta = 52.2°$ for a prism shaped like an equilateral triangle. What is the prism's index of refraction?

55. ‖ Paraxial light rays approach a transparent sphere parallel to an optical axis passing through the center of the sphere. The rays come to a focus on the far surface of the sphere. What is the sphere's index of refraction?

56. ‖ A 6.0-cm-diameter cubic zirconia sphere has an air bubble exactly in the center. As you look into the sphere, how far beneath the surface does the bubble appear to be?

57. ‖ A 1.0-cm-tall object is 10 cm in front of a converging lens that has a 30 cm focal length.

a. Use ray tracing to find the position and height of the image. To do this accurately, use a ruler or paper with a grid. Determine the image distance and image height by making measurements on your diagram.

b. Calculate the image position and height. Compare with your ray-tracing answers in part a.

58. ‖ A 2.0-cm-tall object is 40 cm in front of a converging lens that has a 20 cm focal length.

a. Use ray tracing to find the position and height of the image. To do this accurately, use a ruler or paper with a grid. Determine the image distance and image height by making measurements on your diagram.

b. Calculate the image position and height. Compare with your ray-tracing answers in part a.

59. ‖ A 1.0-cm-tall object is 75 cm in front of a converging lens that has a 30 cm focal length.

a. Use ray tracing to find the position and height of the image. To do this accurately, use a ruler or paper with a grid. Determine the image distance and image height by making measurements on your diagram.

b. Calculate the image position and height. Compare with your ray-tracing answers in part a.

60. ‖ A 2.0-cm-tall object is 15 cm in front of a converging lens that has a 20 cm focal length.

a. Use ray tracing to find the position and height of the image. To do this accurately, use a ruler or paper with a grid. Determine the image distance and image height by making measurements on your diagram.

b. Calculate the image position and height. Compare with your ray-tracing answers in part a.

61. ‖ A 1.0-cm-tall object is 60 cm in front of a diverging lens that has a -30 cm focal length.

a. Use ray tracing to find the position and height of the image. To do this accurately, use a ruler or paper with a grid. Determine the image distance and image height by making measurements on your diagram.

b. Calculate the image position and height. Compare with your ray-tracing answers in part a.

62. ‖ A 2.0-cm-tall object is 15 cm in front of a diverging lens that has a -20 cm focal length.

a. Use ray tracing to find the position and height of the image. To do this accurately, use a ruler or paper with a grid.

Determine the image distance and image height by making measurements on your diagram.

b. Calculate the image position and height. Compare with your ray-tracing answers in part a.

63. ‖ To determine the focal length of a lens, you place the lens in front of a small lightbulb and then adjust a viewing screen to get a sharply focused image. Varying the lens position produces the following data:

Bulb to lens (cm)	Lens to screen (cm)
20	61
22	47
24	39
26	37
28	32

Use the best-fit line of an appropriate graph to determine the focal length of the lens.

64. ‖ A 1.0-cm-tall object is 20 cm in front of a concave mirror that has a 60 cm focal length. Calculate the position and height of the image. State whether the image is in front of or behind the mirror, and whether the image is upright or inverted.

65. ‖ A 1.0-cm-tall object is 20 cm in front of a convex mirror that has a -60 cm focal length. Calculate the position and height of the image. State whether the image is in front of or behind the mirror, and whether the image is upright or inverted.

66. ‖ BIO The illumination lights in an operating room use a concave mirror to focus an image of a bright lamp onto the surgical site. One such light uses a mirror with a 30 cm radius of curvature. If the mirror is 1.2 m from the patient, how far should the lamp be from the mirror?

67. ‖ BIO A dentist uses a curved mirror to view the back side of teeth in the upper jaw. Suppose she wants an upright image with a magnification of 1.5 when the mirror is 1.2 cm from a tooth. Should she use a convex or a concave mirror? What focal length should it have?

68. ‖ A 2.0-cm-tall candle flame is 2.0 m from a wall. You happen to have a lens with a focal length of 32 cm. How many places can you put the lens to form a well-focused image of the candle flame on the wall? For each location, what are the height and orientation of the image?

69. ‖ A lightbulb is 3.0 m from a wall. What are the focal length and the position (measured from the bulb) of a lens that will form an image on the wall that is twice the size of the lightbulb?

70. ‖ BIO a. Estimate the diameter of your eyeball.

b. Bring this page up to the closest distance at which the text is sharp—not the closest at which you can still read it, but the closest at which the letters remain sharp. If you wear glasses or contact lenses, leave them on. This distance is called the *near point* of your (possibly corrected) eye. Measure it.

c. Estimate the effective focal length of your eye. The effective focal length includes the focusing due to the lens, the curvature of the cornea, and any corrections you wear. Ignore the effects of the fluid in your eye.

71. ‖ A slide projector needs to create a 98-cm-high image of a 2.0-cm-tall slide. The screen is 300 cm from the slide.

a. What focal length does the lens need? Assume that it is a thin lens.

b. How far should you place the lens from the slide?

72. ‖ A lens placed 10 cm in front of an object creates an upright image twice the height of the object. The lens is then moved along the optical axis until it creates an inverted image twice the height of the object. How far did the lens move?

73. ‖ An object is 60 cm from a screen. What are the radii of a symmetric converging plastic lens (i.e., two equally curved surfaces) that will form an image on the screen twice the height of the object?

74. ‖ A sports photographer has a 150-mm-focal-length lens on his camera. The photographer wants to photograph a sprinter running straight away from him at 5.0 m/s. What is the speed (in mm/s) of the sprinter's image at the instant the sprinter is 10 m in front of the lens?

75. ‖ A concave mirror has a 40 cm radius of curvature. How far from the mirror must an object be placed to create an upright image three times the height of the object?

76. ‖‖ A 2.0-cm-tall object is placed in front of a mirror. A 1.0-cm-tall upright image is formed behind the mirror, 150 cm from the object. What is the focal length of the mirror?

77. ‖ A spherical mirror of radius R has its center at C, as shown in FIGURE P23.77. A ray parallel to the axis reflects through F, the focal point. Prove that $f = R/2$ if $\phi \ll 1$ rad.

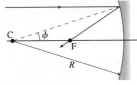

FIGURE P23.77

Challenge Problems

78. Consider a lens having index of refraction n_2 and surfaces with radii R_1 and R_2. The lens is immersed in a fluid that has index of refraction n_1.
    a. Derive a generalized lens' maker's equation to replace Equation 23.27 when the lens is surrounded by a medium other than air. That is, when $n_1 \neq 1$.
    b. A symmetric converging glass lens (i.e., two equally curved surfaces) has two surfaces with radii of 40 cm. Find the focal length of this lens in air and the focal length of this lens in water.

79. FIGURE CP23.79 shows a light ray that travels from point A to point B. The ray crosses the boundary at position x, making angles θ_1 and θ_2 in the two media. Suppose that you did *not* know Snell's law.
    a. Write an expression for the *time t* it takes the light ray to travel from A to B. Your expression should be in terms of the distances a, b, and w; the variable x; and the indices of refraction n_1 and n_2.

    b. The time depends on x. There's one value of x for which the light travels from A to B in the shortest possible time. We'll call it x_{min}. Write an expression (but don't try to solve it!) from which x_{min} could be found.
    c. Now, by using the geometry of the figure, derive Snell's law from your answer to part b.
    You've proven that Snell's law is equivalent to the statement that "light traveling between two points follows the path that requires the shortest time." This interesting way of thinking about refraction is called *Fermat's principle*.

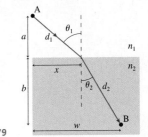

FIGURE P23.79

80. A fortune teller's "crystal ball" (actually just glass) is 10 cm in diameter. Her secret ring is placed 6.0 cm from the edge of the ball.
    a. An image of the ring appears on the opposite side of the crystal ball. How far is the image from the center of the ball?
    b. Draw a ray diagram showing the formation of the image.
    c. The crystal ball is removed and a thin lens is placed where the center of the ball had been. If the image is still in the same position, what is the focal length of the lens?

81. A beam of white light enters a transparent material. Wavelengths for which the index of refraction is n are refracted at angle θ_2. Wavelengths for which the index of refraction is $n + \delta n$, where $\delta n \ll n$, are refracted at angle $\theta_2 + \delta\theta$.
    a. Show that the angular separation in radians is $\delta\theta = -(\delta n/n)\tan\theta_2$.
    b. A beam of white light is incident on a piece of glass at 30.0°. Deep violet light is refracted 0.28° more than deep red light. The index of refraction for deep red light is known to be 1.552. What is the index of refraction for deep violet light?

82. Consider an object of thickness ds (parallel to the axis) in front of a lens or mirror. The image of the object has thickness ds'. Define the *longitudinal magnification* as $M = ds'/ds$. Prove that $M = -m^2$, where m is the lateral magnification.

STOP TO THINK ANSWERS

Stop to Think 23.1: c. The light spreads vertically as it goes through the vertical aperture. The light spreads horizontally due to different points on the horizontal lightbulb.

Stop to Think 23.2: c. There's one image behind the vertical mirror and a second behind the horizontal mirror. A third image in the corner arises from rays that reflect twice, once off each mirror.

Stop to Think 23.3: a. The ray travels closer to the normal in both media 1 and 3 than in medium 2, so n_1 and n_3 are both larger than n_2. The angle is smaller in medium 3 than in medium 1, so $n_3 > n_1$.

Stop to Think 23.4: e. The rays from the object are diverging. Without a lens, the rays cannot converge to form any kind of image on the screen.

Stop to Think 23.5: a, e, or f. Any of these will increase the angle of refraction θ_2.

Stop to Think 23.6: Away from. You need to decrease s' to bring the image plane onto the screen. s' is decreased by increasing s.

Stop to Think 23.7: c. A concave mirror forms a real image in front of the mirror. Because the object distance is $s \approx \infty$, the image distance is $s' \approx f$.

24 Optical Instruments

The world's greatest collection of telescopes is on the summit of Mauna Kea on the Big Island of Hawaii, towering 4200 m (13,800 ft) over the Pacific Ocean.

▶ **Looking Ahead** The goal of Chapter 24 is to understand some common optical instruments and their limitations.

Lenses in Combination

The "lenses" of optical instruments are always built with several individual lenses to give better optical performance.

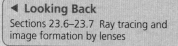

A cross section of a typical camera lens shows that it is built of 5 individual lenses and an adjustable iris.

You'll learn how to analyze a system with multiple lenses.

Optical Systems That Magnify

Lenses and mirrors can be used to magnify objects both near and far. Optical instruments open a realm far beyond what the unaided eye can see.

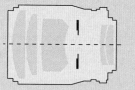

A simple magnifying glass has a low magnification of only 2× or 3×.

A microscope uses two sets of lenses in combination to produce magnifications of up to 1000×.

Small telescopes use lenses; larger telescopes use a curved mirror as the primary optical element.

The Camera

A camera uses a lens to project a real image onto a light-sensitive detector.

Although a modern digital camera is very complex, at its heart it's just a light-tight box with a lens to focus the image.

You'll learn about focusing, zoom, and exposure.

◀ **Looking Back**
Sections 23.6–23.7 Ray tracing and image formation by lenses

The Human Eye

The human eye is much like a camera: The cornea and lens together focus a real image onto the retina.

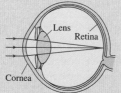

You'll discover how eyeglasses and contact lenses are used to correct defects of vision.

Resolution of Lenses

Light passing through a lens undergoes diffraction, just like light passing through a circular hole. Diffraction limits a lens's ability to form a perfectly focused image.

An ideal lens would have focused the light to two points. Instead, we get two overlapped diffraction patterns.

You'll learn about *Rayleigh's criterion* for when two images can be resolved.

◀ **Looking Back**
Section 22.5 Circular diffraction

24.1 Lenses in Combination

Only the simplest magnifiers are built with a single lens of the sort we analyzed in Chapter 23. Optical instruments, such as microscopes and cameras, are invariably built with multiple lenses. The reason, as we'll see, is to improve the image quality.

The analysis of multi-lens systems requires only one new rule: **The image of the first lens acts as the object for the second lens.** To see why this is so, **FIGURE 24.1** shows a simple telescope consisting of a large-diameter converging lens, called the *objective,* and a smaller converging lens used as the *eyepiece.* (We'll analyze telescopes more thoroughly later in the chapter.) Highlighted are the three special rays you learned to use in Chapter 23:

- A ray parallel to the optical axis refracts through the focal point.
- A ray through the focal point refracts parallel to the optical axis.
- A ray through the center of the lens is undeviated.

FIGURE 24.1 Ray-tracing diagram of a simple astronomical telescope.

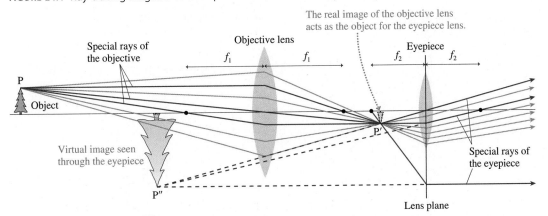

The rays passing through the objective converge to a real image at P′, but they don't stop there. Instead, light rays *diverge* from P′ as they approach the second lens. **As far as the eyepiece is concerned, the rays are coming from P′, and thus P′ acts as the object for the second lens.** The three special rays passing through the objective lens are sufficient to locate the image P′, but these rays are generally *not* the special rays for the second lens. However, other rays converging at P′ leave at the correct angles to be the special rays for the eyepiece. That is, a new set of special rays is drawn from P′ to the second lens and used to find the final image point P″.

NOTE ▶ One ray seems to "miss" the eyepiece lens, but this isn't a problem. All rays passing through the lens converge to (or diverge from) a single point, and the purpose of the special rays is to locate that point. To do so, we can let the special rays refract as they cross the *lens plane,* regardless of whether the physical lens really extends that far. ◀

EXAMPLE 24.1 **A camera lens**

The "lens" on a camera is usually a combination of two or more single lenses. Consider a camera in which light passes first through a diverging lens, with $f_1 = -120$ mm, then a converging lens, with $f_2 = 42$ mm, spaced 60 mm apart. A reasonable definition of the *effective focal length* of this lens combination is the focal length of a *single* lens that could produce an image in the same location if placed at the midpoint of the lens combination. A 10-cm-tall object is 500 mm from the first lens.

a. What are the location, size, and orientation of the image?
b. What is the effective focal length of the double-lens system used in this camera?

MODEL Each lens is a thin lens. The image of the first lens is the object for the second.

VISUALIZE The ray-tracing diagram of **FIGURE 24.2** shows the production of a real, inverted image ≈ 55 mm behind the second lens.

Continued

SOLVE

a. $s_1 = 500$ mm is the object distance of the first lens. Its image, a virtual image, is found from the thin-lens equation:

$$\frac{1}{s_1'} = \frac{1}{f_1} - \frac{1}{s_1} = \frac{1}{-120 \text{ mm}} - \frac{1}{500 \text{ mm}} = -0.0103 \text{ mm}^{-1}$$

$$s_1' = -97 \text{ mm}$$

This is consistent with the ray-tracing diagram. The image of the first lens now acts as the object for the second lens. Because the lenses are 60 mm apart, the object distance is $s_2 = 97$ mm $+ 60$ mm $= 157$ mm. A second application of the thin-lens equation yields

$$\frac{1}{s_2'} = \frac{1}{f_2} - \frac{1}{s_2} = \frac{1}{42 \text{ mm}} - \frac{1}{157 \text{ mm}} = 0.0174 \text{ mm}^{-1}$$

$$s_2' = 57 \text{ mm}$$

The image of the lens combination is 57 mm behind the second lens. The lateral magnifications of the two lenses are

$$m_1 = -\frac{s_1'}{s_1} = -\frac{-97 \text{ cm}}{500 \text{ cm}} = 0.194$$

$$m_2 = -\frac{s_2'}{s_2} = -\frac{57 \text{ cm}}{157 \text{ cm}} = -0.363$$

The second lens magnifies the image of the first lens, which magnifies the object, so **the total magnification is the product of the individual magnifications:**

$$m = m_1 m_2 = -0.070$$

Thus the image is 57 mm behind the second lens, inverted (m is negative), and 0.70 cm tall.

b. If a single lens midway between these two lenses produced an image in the same plane, its object and image distances would be $s = 500$ mm $+ 30$ mm $= 530$ mm and $s' = 57$ mm $+ 30$ mm $= 87$ mm. A final application of the thin-lens equation gives the effective focal length:

$$\frac{1}{f_{\text{eff}}} = \frac{1}{s} + \frac{1}{s'} = \frac{1}{530 \text{ mm}} + \frac{1}{87 \text{ mm}} = 0.0134 \text{ mm}^{-1}$$

$$f_{\text{eff}} = 75 \text{ mm}$$

ASSESS This combination lens would be sold as a "75 mm lens."

FIGURE 24.2 Pictorial representation of a combination lens.

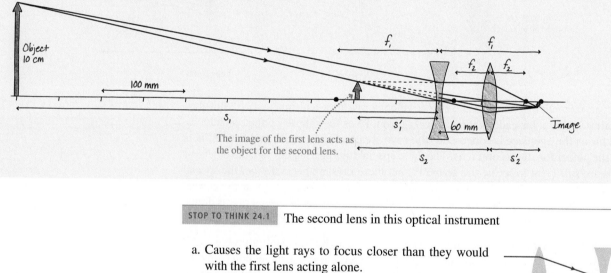

STOP TO THINK 24.1 The second lens in this optical instrument

a. Causes the light rays to focus closer than they would with the first lens acting alone.
b. Causes the light rays to focus farther away than they would with the first lens acting alone.
c. Inverts the image but does not change where the light rays focus.
d. Prevents the light rays from reaching a focus.

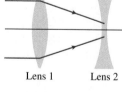

Lens 1   Lens 2

FIGURE 24.3 A camera.

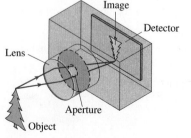

24.2 The Camera

A **camera,** shown in **FIGURE 24.3**, "takes a picture" by using a lens to form a real, inverted image on a light-sensitive detector in a light-tight box. Film was the detector of choice for well over a hundred years, but today's digital cameras use an electronic detector called a *charge-coupled device,* or CCD.

The camera "lens" is always a combination of two or more individual lenses. The simplest such lens, shown in **FIGURE 24.4**, consists of a converging lens and a somewhat weaker diverging lens. This combination of positive and negative lenses corrects some of the defects inherent in single lenses, as we'll discuss later in the chapter. As Example 24.1 suggested, we can model a combination lens as a single lens with an **effective focal length** (usually called simply "the focal length") f. A *zoom lens* changes the effective focal length by changing the spacing between the converging lens and the diverging lens; this is what happens when the lens barrel on your digital camera moves in and out as you use the zoom. A typical digital camera has a lens whose effective focal length can be varied from 6 mm to 18 mm, giving, as we'll see, a 3× zoom.

FIGURE 24.4 A simple camera lens is a combination lens.

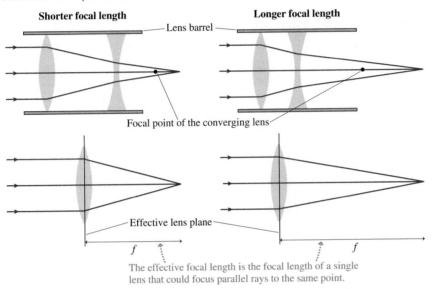

A camera must carry out two important functions: focus the image on the detector and control the exposure. Cameras are focused by moving the lens forward or backward until the image is well focused on the detector. Most modern cameras do this automatically, but older cameras required manual focusing.

EXAMPLE 24.2 | **Focusing a camera**

Your digital camera lens, with an effective focal length of 10.0 mm, is focused on a flower 20.0 cm away. You then turn to take a picture of a distant landscape. How far, and in which direction, must the lens move to bring the landscape into focus?

MODEL Model the camera's combination lens as a single thin lens with $f = 10.0$ mm. Image and object distances are measured from the effective lens plane. Assume all the lenses in the combination move together as the camera refocuses.

SOLVE The flower is at object distance $s = 20.0$ cm $= 200$ mm. When the camera is focused, the image distance between the effective lens plane and the detector is found by solving the thin-lens equation $1/s + 1/s' = 1/f$ to give

$$s' = \left(\frac{1}{f} - \frac{1}{s}\right)^{-1} = \left(\frac{1}{10.0 \text{ mm}} - \frac{1}{200 \text{ mm}}\right)^{-1} = 10.5 \text{ mm}$$

The distant landscape is effectively at object distance $s = \infty$, so its image distance is $s' = f = 10.0$ mm. To refocus as you shift scenes, the lens must move 0.5 mm closer to the detector.

ASSESS The required motion of the lens is very small, about the diameter of the lead used in a mechanical pencil.

Zoom Lenses

For objects more than 10 focal lengths from the lens (roughly $s > 20$ cm for a typical digital camera), the approximation $s \gg f$ (and thus $1/s \ll 1/f$) leads to $s' \approx f$. In other words, objects more than about 10 focal lengths away are essentially "at infinity," and we know that the parallel rays from an infinitely distant object are focused one focal length behind the lens. For such an object, the lateral magnification of the image is

$$m = -\frac{s'}{s} \approx -\frac{f}{s} \tag{24.1}$$

The magnification is much less than 1, because $s \gg f$, so the image on the detector is much smaller than the object itself. This comes as no surprise. More important, **the size of the image is directly proportional to the focal length of the lens.** We saw in Figure 24.4 that the effective focal length of a combination lens is easily changed by varying the distance between the individual lenses, and this is exactly how a zoom lens works. A lens that can be varied from $f_{min} = 6$ mm to $f_{max} = 18$ mm gives magnifications spanning a factor of 3, and that is why you see it specified as a $3\times$ zoom lens.

Controlling the Exposure

The camera also must control the amount of light reaching the detector. Too little light results in photos that are *underexposed;* too much light gives *overexposed* pictures. Both the shutter and the lens diameter help control the exposure.

The *shutter* is "opened" for a selected amount of time as the image is recorded. Older cameras used a spring-loaded mechanical shutter that literally opened and closed; digital cameras electronically control the amount of time the detector is active. Either way, the exposure—the amount of light captured by the detector—is directly proportional to the time the shutter is open. Typical exposure times range from 1/1000 s or less for a sunny scene to 1/30 s or more for dimly lit or indoor scenes. The exposure time is generally referred to as the *shutter speed.*

The amount of light passing through the lens is controlled by an adjustable **aperture,** also called an *iris* because it functions much like the iris of your eye. The aperture sets the effective diameter D of the lens. The full area of the lens is used when the aperture is fully open, but a *stopped-down* aperture allows light to pass through only the central portion of the lens.

The light intensity on the detector is directly proportional to the area of the lens; a lens with twice as much area will collect and focus twice as many light rays from the object to make an image twice as bright. The lens area is proportional to the square of its diameter, so the intensity I is proportional to D^2. The light intensity—power per square meter—is also *inversely* proportional to the area of the image. That is, the light reaching the detector is more intense if the rays collected from the object are focused into a small area than if they are spread out over a large area. The lateral size of the image is proportional to the focal length of the lens, as we saw in Equation 24.1, so the *area* of the image is proportional to f^2 and thus I is proportional to $1/f^2$. Altogether, $I \propto D^2/f^2$.

By long tradition, the light-gathering ability of a lens is specified by its **f-number,** defined as

$$f\text{-number} = \frac{f}{D} \tag{24.2}$$

The *f*-number of a lens may be written either as *f*/4.0, to mean that the *f*-number is 4.0, or as F4.0. The instruction manuals with some digital cameras call this the *aperture value* rather than the *f*-number. A digital camera in fully automatic mode does not display shutter speed or *f*-number, but that information is displayed if you set your camera to any of the other modes. For example, the display 1/125 F5.6 means that your camera is going to achieve the correct exposure by adjusting the diameter of the lens aperture to give *f*/D = 5.6 and by opening the shutter for 1/125 s. If your lens's effective focal length is 10 mm, the diameter of the lens aperture will be

$$D = \frac{f}{f\text{-number}} = \frac{10 \text{ mm}}{5.6} = 1.8 \text{ mm}$$

An iris can change the effective diameter of a lens and thus the amount of light reaching the detector.

NOTE ▶ The *f* in *f*-number is not the focal length *f*; it's just a name. And the / in *f*/4 does not mean division; it's just a notation. These both derive from the long history of photography. ◀

Because the aperture diameter is in the denominator of the *f*-number, a *larger-diameter* aperture, which gathers more light and makes a brighter image, has a *smaller f*-number. The light intensity on the detector is related to the lens's *f*-number by

$$I \propto \frac{D^2}{f^2} = \frac{1}{(f\text{-number})^2} \qquad (24.3)$$

Historically, a lens's *f*-numbers could be adjusted in the sequence 2.0, 2.8, 4.0, 5.8, 8.0, 11, 16. Each differs from its neighbor by a factor of $\sqrt{2}$, so changing the lens by one "*f* stop" changed the light intensity by a factor of 2. A modern digital camera is able to adjust the *f*-number continuously.

The exposure, the total light reaching the detector while the shutter is open, depends on the product $I\Delta t_{\text{shutter}}$. A small *f*-number (large aperture diameter *D*) and short $\Delta t_{\text{shutter}}$ can produce the same exposure as a larger *f*-number (smaller aperture) and a longer $\Delta t_{\text{shutter}}$. It might not make any difference for taking a picture of a distant mountain, but action photography needs very short shutter times to "freeze" the action. Thus action photography requires a large-diameter lens with a small *f*-number.

Focal length and *f*-number information is stamped on a camera lens. This lens is labeled 5.8–23.2 mm 1:2.6–5.5. The first numbers are the range of focal lengths. They span a factor of 4, so this is a 4× zoom lens. The second numbers show that the minimum *f*-number ranges from *f*/2.6 (for the *f* = 5.8 mm focal length) to *f*/5.5 (for the *f* = 23.2 mm focal length).

EXAMPLE 24.3 **Capturing the action**

Before a race, a photographer finds that she can make a perfectly exposed photo of the track while using a shutter speed of 1/250 s and a lens setting of *f*/8.0. To freeze the sprinters as they go past, she plans to use a shutter speed of 1/1000 s. To what *f*-number must she set her lens?

MODEL The exposure depends on $I\Delta t_{\text{shutter}}$, and the light intensity depends inversely on the square of the *f*-number.

SOLVE Changing the shutter speed from 1/250 s to 1/1000 s will reduce the light reaching the detector by a factor of 4. To compensate, she needs to let 4 times as much light through the lens. Because $I \propto 1/(f\text{-number})^2$, the intensity will increase by a factor of 4 if she *decreases* the *f*-number by a factor of 2. Thus the correct lens setting is *f*/4.0.

ASSESS To keep the photo properly exposed, a decreased shutter time must be balanced by an increased lens aperture diameter.

The Detector

For traditional cameras, the light-sensitive detector is film. Today's digital cameras use an electronic light-sensitive surface called a *charge-coupled device* or **CCD**. A CCD consists of a rectangular array of many millions of small detectors called **pixels.** When light hits one of these pixels, it generates an electric charge proportional to the light intensity. Thus an image is recorded on the CCD in terms of little packets of charge. After the CCD has been exposed, the charges are read out, the signal levels are digitized, and the picture is stored in the digital memory of the camera.

FIGURE 24.5a shows a CCD "chip" and, schematically, the magnified appearance of the pixels on its surface. To record color information, different pixels are covered by red, green, or blue filters. A pixel covered by a green filter, for instance, records only the intensity of the green light hitting it. Later, the camera's microprocessor interpolates nearby colors to give each pixel an overall true color. The pixels are so small that the picture looks "smooth" even after some enlargement, but, as you can see in **FIGURE 24.5b**, sufficient magnification reveals the individual pixels.

FIGURE 24.5 The CCD detector used in a digital camera.

(a) 2500 × 2000 pixels

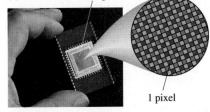

1 pixel

(b)

STOP TO THINK 24.2 A photographer has adjusted his camera for a correct exposure with a short-focal-length lens. He then decides to zoom in by increasing the focal length. To maintain a correct exposure without changing the shutter speed, the diameter of the lens aperture should

   a. Be increased.   b. Be decreased.   c. Stay the same.

24.3 Vision

The human eye is a marvelous and intricate organ. If we leave the biological details to biologists and focus on the eye's optical properties, we find that it functions very much like a camera. Like a camera, the eye has refracting surfaces that focus incoming light rays, an adjustable iris to control the light intensity, and a light-sensitive detector.

FIGURE 24.6 shows the basic structure of the eye. It is roughly spherical, about 2.4 cm in diameter. The transparent **cornea,** which is somewhat more sharply curved, and the *lens* are the eye's refractive elements. The eye is filled with a clear, jellylike fluid called the *aqueous humor* (in front of the lens) and the *vitreous humor* (behind the lens). The indices of refraction of the aqueous and vitreous humors are 1.34, only slightly different from water. The lens, although not uniform, has an average index of 1.44. The **pupil,** a variable-diameter aperture in the **iris,** automatically opens and closes to control the light intensity. A fully dark-adapted eye can open to $\approx$ 8 mm, and the pupil closes down to $\approx$ 1.5 mm in bright sun. This corresponds to f-numbers from roughly $f/3$ to $f/16$, very similar to a camera.

FIGURE 24.6 The human eye.

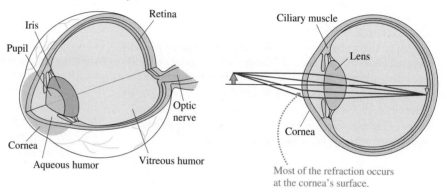

Most of the refraction occurs at the cornea's surface.

FIGURE 24.7 Wavelength sensitivity of the three types of cones in the human retina.

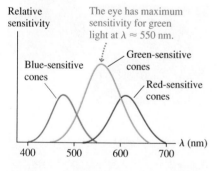

The eye's detector, the **retina,** consists of specialized light-sensitive cells called *rods* and *cones*. The rods, sensitive mostly to light and dark, are most important in very dim lighting. Color vision, which requires somewhat more light, is due to the cones, of which there are three types. FIGURE 24.7 shows the wavelength responses of the cones. They have overlapping ranges, especially the red- and green-sensitive cones, so two or even all three cones respond to light of any particular wavelength. The relative response of the different cones is interpreted by your brain as light of a particular color. Color is a *perception,* a response of our sensory and nervous systems, not something inherent in the light itself. Other animals, with slightly different retinal cells, can see ultraviolet or infrared wavelengths that we cannot see.

Focusing and Accommodation

The eye, like a camera, focuses light rays to an inverted image on the retina. Perhaps surprisingly, most of the refractive power of the eye is due to the cornea, not the lens. The cornea is a sharply curved, spherical surface, and you learned in Chapter 23 that images are formed by refraction at a spherical surface. The rather large difference between the index of refraction of air and that of the aqueous humor causes a significant refraction of light rays at the cornea. In contrast, there is much less difference between the indices of the lens and its surrounding fluid, so refraction at the lens surfaces is weak. The lens is important for fine-tuning, but the air-cornea boundary is responsible for the majority of the refraction.

You can recognize the power of the cornea if you open your eyes underwater. Everything is very blurry! When light enters the cornea through water, rather than through air, there's almost no difference in the indices of refraction at the surface. Light rays pass through the cornea with almost no refraction, so what little focusing ability you have while underwater is due to the lens alone.

A camera focuses by moving the lens. The eye focuses by changing the focal length of the lens, a feat it accomplishes by using the *ciliary muscles* to change the curvature of the lens surface. The ciliary muscles are relaxed when you look at a distant scene. Thus the lens surface is relatively flat and the lens has its longest focal length. As you shift your gaze to a nearby object, the ciliary muscles contract and cause the lens to bulge. This process, called **accommodation,** decreases the lens's radius of curvature and thus decreases its focal length.

The farthest distance at which a relaxed eye can focus is called the eye's **far point** (FP). The far point of a normal eye is infinity; that is, the eye can focus on objects extremely far away. The closest distance at which an eye can focus, using maximum accommodation, is the eye's **near point** (NP). (Objects can be *seen* closer than the near point, but they're not sharply focused on the retina.) Both situations are shown in FIGURE 24.8.

Vision Defects and Their Correction

The near point of normal vision is considered to be 25 cm, but the near point of any individual changes with age. The near point of young children can be as little as 10 cm. The "normal" 25 cm near point is characteristic of young adults, but the near point of most individuals begins to move outward by age 40 or 45 and can reach 200 cm by age 60. This loss of accommodation, which arises because the lens loses flexibility, is called **presbyopia.** Even if their vision is otherwise normal, individuals with presbyopia need reading glasses to bring their near point back to 25 or 30 cm, a comfortable distance for reading.

Presbyopia is known as a *refractive error* of the eye. Two other common refractive errors are *hyperopia* and *myopia.* All three can be corrected with lenses—either eyeglasses or contact lenses—that assist the eye's focusing. Corrective lenses are prescribed not by their focal length but by their **power.** The power of a lens is the inverse of its focal length:

$$\text{Power of a lens} = P = \frac{1}{f} \tag{24.4}$$

A lens with more power (shorter focal length) causes light rays to refract through a larger angle. The SI unit of lens power is the **diopter,** abbreviated D, defined as $1 \text{ D} = 1 \text{ m}^{-1}$. Thus a lens with $f = 50 \text{ cm} = 0.50 \text{ m}$ has power $P = 2.0 \text{ D}$.

A person who is *farsighted* can see faraway objects (but even then must use some accommodation rather than a relaxed eye), but his near point is larger than 25 cm, often much larger, so he cannot focus on nearby objects. The cause of farsightedness—called **hyperopia**—is an eyeball that is too short for the refractive power of the cornea and lens. As FIGURES 24.9a and b on the next page show, no amount of accommodation allows the eye to focus on an object 25 cm away, the normal near point.

With hyperopia, the eye needs assistance to focus the rays from a near object onto the closer-than-normal retina. This assistance is obtained by adding refractive power with the positive (i.e., converging) lens shown in FIGURE 24.9c. To understand why this works, recall that the image of a first lens acts as the object for a second lens. The goal is to allow the person to focus on an object 25 cm away. If a corrective lens forms an upright, virtual image at the person's actual near point, that virtual image acts as an object for the eye itself and, with maximum accommodation, the eye can focus these rays onto the retina. Presbyopia, the loss of accommodation with age, is corrected in the same way.

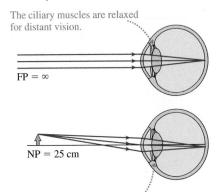

FIGURE 24.8 Normal vision of far and near objects.

The ciliary muscles are relaxed for distant vision.

FP = ∞

NP = 25 cm

The ciliary muscles are contracted for near vision, causing the lens to curve more.

The optometrist's prescription is −2.25 D for the right eye (top) and −2.50 D for the left (bottom), the minus sign indicating that these are diverging lenses. The optometrist doesn't write the D because the lens maker already knows that prescriptions are in diopters. Most people's eyes are not exactly the same, so each eye usually gets a different lens.

FIGURE 24.9 Hyperopia.

FIGURE 24.10 Myopia.

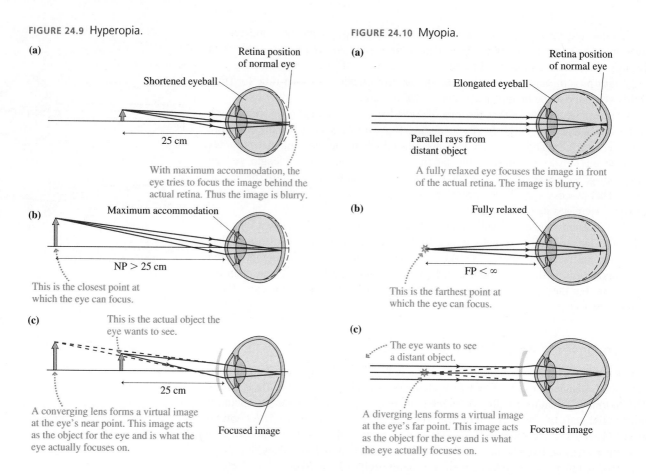

NOTE ▶ Figures 24.9 and 24.10 show the corrective lenses as they are actually shaped—called *meniscus lenses*—rather than with our usual lens shape. Nonetheless, the lens in Figure 24.9c is a converging lens because it's thicker in the center than at the edges. The lens in Figure 24.10c is a diverging lens because it's thicker at the edges than in the center. ◀

A person who is *nearsighted* can clearly see nearby objects when the eye is relaxed (and extremely close objects by using accommodation), but no amount of relaxation allows her to see distant objects. Nearsightedness—called **myopia**—is caused by an eyeball that is too long. As FIGURE 24.10a shows, rays from a distant object come to a focus in front of the retina and have begun to diverge by the time they reach the retina. The eye's far point, shown in FIGURE 24.10b, is less than infinity.

To correct myopia, we needed a diverging lens, as shown in FIGURE 24.10c, to slightly defocus the rays and move the image point back to the retina. To focus on a very distant object, the person needs a corrective lens that forms an upright, virtual image at her actual far point. That virtual image acts as an object for the eye itself and, when fully relaxed, the eye can focus these rays onto the retina.

EXAMPLE 24.4 | **Correcting hyperopia**

Sanjay has hyperopia. The near point of his left eye is 150 cm. What prescription lens will restore normal vision?

MODEL Normal vision will allow Sanjay to focus on an object 25 cm away. In measuring distances, we'll ignore the small space between the lens and his eye.

SOLVE Because Sanjay can see objects at 150 cm, using maximum accommodation, we want a lens that creates a virtual image

at position $s' = -150$ cm (negative because it's a virtual image) of an object held at $s = 25$ cm. From the thin-lens equation,

$$\frac{1}{f} = \frac{1}{s} + \frac{1}{s'} = \frac{1}{0.25 \text{ m}} + \frac{1}{-1.50 \text{ m}} = 3.3 \text{ m}^{-1}$$

$1/f$ is the lens power, and m^{-1} are diopters. Thus the prescription is for a lens with power $P = 3.3$ D.

ASSESS Hyperopia is always corrected with a converging lens.

EXAMPLE 24.5 **Correcting myopia**

Martina has myopia. The far point of her left eye is 200 cm. What prescription lens will restore normal vision?

MODEL Normal vision will allow Martina to focus on a very distant object. In measuring distances, we'll ignore the small space between the lens and her eye.

SOLVE Because Martina can see objects at 200 cm with a fully relaxed eye, we want a lens that will create a virtual image at position $s' = -200$ cm (negative because it's a virtual image) of a distant object at $s = \infty$ cm. From the thin-lens equation,

$$\frac{1}{f} = \frac{1}{s} + \frac{1}{s'} = \frac{1}{\infty \text{ m}} + \frac{1}{-2.0 \text{ m}} = -0.5 \text{ m}^{-1}$$

Thus the prescription is for a lens with power $P = -0.5$ D.

ASSESS Myopia is always corrected with a diverging lens.

STOP TO THINK 24.3 You need to improvise a magnifying glass to read some very tiny print. Should you borrow the eyeglasses from your hyperopic friend or from your myopic friend?

a. The hyperopic friend
c. Either will do.

b. The myopic friend
d. Neither will work.

24.4 Optical Systems That Magnify

The camera, with its fast shutter speed, allows us to capture images of events that take place too quickly for our unaided eye to resolve. Another use of optical systems is to magnify—to see objects smaller or closer together than our eye can see.

The easiest way to magnify an object requires no extra optics at all; simply get closer! The closer you get, the bigger the object appears. Obviously the actual size of the object is unchanged as you approach it, so what exactly is getting "bigger"? Consider the green arrow in FIGURE 24.11a. We can determine the size of its image on the retina by tracing the ray that is undeviated as it passes through the center of a lens. (Here we're modeling the eye's optical system as one thin lens.) If we get closer to the arrow, now shown as red, we find the arrow makes a larger image on the retina. Our brain interprets the larger image as a larger-appearing object. The object's actual size doesn't change, but its *apparent size* gets larger as it gets closer.

Technically, we say that closer objects look larger because they subtend a larger angle θ, called the **angular size** of the object. The red arrow has a larger angular size than the green arrow, $\theta_2 > \theta_1$, so the red arrow looks larger and we can see more detail. But you can't keep increasing an object's angular size because you can't focus on the object if it's closer than your near point, which we'll take to be a normal 25 cm. FIGURE 24.11b defines the angular size θ_{NP} of an object at your near point. If the object's height is h and if we assume the small-angle approximation $\tan\theta \approx \theta$, the maximum angular size viewable by your unaided eye is

$$\theta_{NP} = \frac{h}{25 \text{ cm}} \tag{24.5}$$

Suppose we view the same object, of height h, through the single converging lens in FIGURE 24.12 on the next page. If the object's distance from the lens is less than the lens's focal length, we'll see an enlarged, upright image. Used in this way, the lens is called a **magnifier** or *magnifying glass*. The eye sees the virtual image subtending angle θ, and it can focus on this virtual image as long as the image distance is more than 25 cm. Within the small-angle approximation, the image subtends angle $\theta = h/s$. In practice, we usually want the image to be at distance $s' \approx \infty$ so that we can view it with a relaxed eye as a "distant object." This will be true if the object is very near the focal point: $s \approx f$. In this case, the image subtends angle

$$\theta = \frac{h}{s} \approx \frac{h}{f} \tag{24.6}$$

FIGURE 24.11 Angular size.

(a) Same object at two different distances

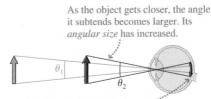

As the object gets closer, the angle it subtends becomes larger. Its *angular size* has increased.

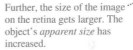

Further, the size of the image on the retina gets larger. The object's *apparent size* has increased.

(b)

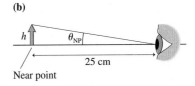

Near point

FIGURE 24.12 The magnifier.

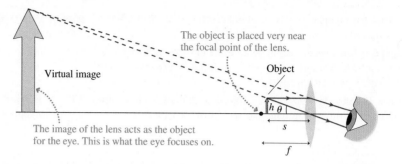

The object is placed very near the focal point of the lens.

Virtual image

Object

The image of the lens acts as the object for the eye. This is what the eye focuses on.

Let's define the **angular magnification** M as

$$M = \frac{\theta}{\theta_{\text{NP}}} \tag{24.7}$$

Angular magnification is the increase in the *apparent size* of the object that you achieve by using a magnifying lens rather than simply holding the object at your near point. Substituting from Equations 24.5 and 24.6, we find the angular magnification of a magnifying glass is

$$M = \frac{25 \text{ cm}}{f} \tag{24.8}$$

The angular magnification depends on the focal length of the lens but not on the size of the object. Although it would appear we could increase angular magnification without limit by using lenses with shorter and shorter focal lengths, the inherent limitations of lenses we discuss in the next section limit the magnification of a simple lens to about 4×. Slightly more complex magnifiers with two lenses reach 20×, but beyond that one would use a microscope.

> NOTE ▶ Don't confuse angular magnification with lateral magnification. Lateral magnification m compares the height of an object to the height of its image. The lateral magnification of a magnifying glass is $\approx \infty$ because the virtual image is at $s' \approx \infty$, but that doesn't make the object seem infinitely big. Its apparent size is determined by the angle subtended on your retina, and that angle remains finite. Thus angular magnification tells us how much bigger things appear. ◀

The Microscope

A microscope, whose major parts are shown in FIGURE 24.13a, can attain a magnification of up to 1000× by a *two-step* magnification process. A specimen to be observed is placed on the *stage* of the microscope, directly beneath the **objective,** a converging lens with a relatively short focal length. The objective creates a magnified real image that is further enlarged by the **eyepiece.** Both the objective and the eyepiece are complex combination lenses, but we'll model them as single thin lenses. It's common for a prism to bend the rays so that the eyepiece is at a comfortable viewing angle. However, we'll consider a simplified version of a microscope in which the light travels along a straight tube.

FIGURE 24.13b shows the optics in more detail. The object is placed just outside the focal point of the objective, which then creates a highly magnified real image with lateral magnification $m = -s'/s$. The object is so close to the focal point that $s \approx f_{\text{obj}}$ is an excellent approximation. In addition, the focal lengths of the objective and the eyepiece are much less than the tube length L, so $s' \approx L$ is another good approximation. With these approximations, the lateral magnification of the objective is

$$m_{\text{obj}} = -\frac{s'}{s} \approx -\frac{L}{f_{\text{obj}}} \tag{24.9}$$

FIGURE 24.13 The microscope.

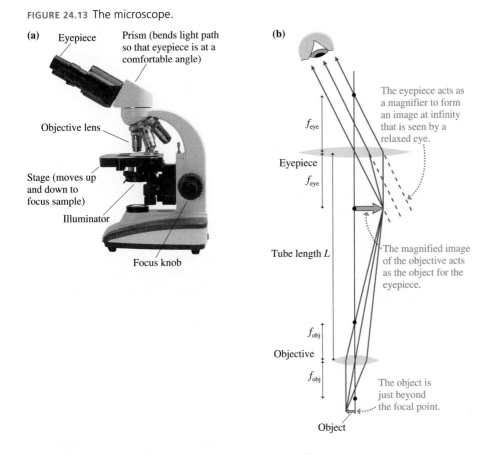

(a)
Eyepiece

Prism (bends light path so that eyepiece is at a comfortable angle)

Objective lens

Stage (moves up and down to focus sample)

Illuminator

Focus knob

(b)

The eyepiece acts as a magnifier to form an image at infinity that is seen by a relaxed eye.

f_{eye}

Eyepiece

f_{eye}

Tube length L

The magnified image of the objective acts as the object for the eyepiece.

f_{obj}

Objective

f_{obj}

The object is just beyond the focal point.

Object

The image of the objective acts as the object for the eyepiece, which functions as a simple magnifier. The angular magnification of the eyepiece is given by Equation 24.8, $M_{eye} = (25 \text{ cm})/f_{eye}$. Together, the objective and eyepiece produce a total angular magnification

$$M = m_{obj}M_{eye} = -\frac{L}{f_{obj}}\frac{25 \text{ cm}}{f_{eye}} \qquad (24.10)$$

The minus sign shows that the image seen in a microscope is inverted.

In practice, the magnifications of the objective (without the minus sign) and the eyepiece are stamped on the barrels. A set of objectives on a rotating turret might include 10×, 20×, 40×, and 100×. When combined with a 10× eyepiece, the microscope's total angular magnification ranges from 100× to 1000×. In addition, most biological microscopes are standardized with a tube length $L = 160$ mm. Thus a 40× objective has focal length $f_{obj} = 160 \text{ mm}/40 = 4.0$ mm.

EXAMPLE 24.6 **Viewing blood cells**

A pathologist inspects a sample of 7-μm-diameter human blood cells under a microscope. She selects a 40× objective and a 10× eyepiece. What size object, viewed from 25 cm, has the same apparent size as a blood cell seen through the microscope?

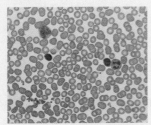

MODEL Angular magnification compares the magnified angular size to the angular size seen at the near-point distance of 25 cm.

SOLVE The microscope's angular magnification is $M = -(40) \times (10) = -400$. The magnified cells will have the same apparent size as an object $400 \times 7 \ \mu\text{m} \approx 3$ mm in diameter seen from a distance of 25 cm.

ASSESS 3 mm is about the size of a capital O in this textbook, so a blood cell seen through the microscope will have about the same apparent size as an O seen from a comfortable reading distance.

A biologist rotates the turret of a microscope to replace a $20\times$ objective with a $10\times$ objective. To keep the same overall magnification, the focal length of the eyepiece must be

a. Doubled.   b. Halved.   c. Kept the same.
d. The magnification cannot be kept the same if the objective is changed.

The Telescope

A microscope magnifies small, nearby objects to look large. A telescope magnifies distant objects, which might be quite large, so that we can see details that are blended together when seen by eye.

FIGURE 24.14 shows the optical layout of a simple telescope. A large-diameter objective lens (larger lenses collect more light and thus can see fainter objects) collects the parallel rays from a distant object ($s = \infty$) and forms a real, inverted image at distance $s' = f_{obj}$. Unlike a microscope, which uses a short-focal-length objective, the focal length of a telescope objective is very nearly the length of the telescope tube. Then, just as in the microscope, the eyepiece functions as a simple magnifier. The viewer observes an inverted image, but that's not a serious problem in astronomy. Terrestrial telescopes use a different design to obtain an upright image.

FIGURE 24.14 A refracting telescope.

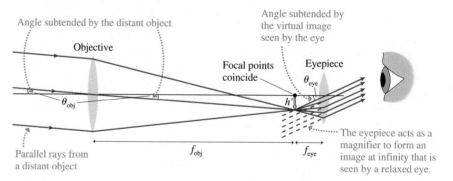

Suppose the distant object, as seen by the objective lens, subtends angle θ_{obj}. If the image seen through the eyepiece subtends a larger angle θ_{eye}, then the angular magnification is $M = \theta_{eye}/\theta_{obj}$. We can see from the undeviated ray passing through the center of the objective lens that (using the small-angle approximation)

$$\theta_{obj} \approx -\frac{h'}{f_{obj}}$$

where the minus sign indicates the inverted image. The image of height h' acts as the object for the eyepiece, and we can see that the final image observed by the viewer subtends angle

$$\theta_{eye} = \frac{h'}{f_{eye}}$$

Consequently, the angular magnification of a telescope is

$$M = \frac{\theta_{eye}}{\theta_{obj}} = -\frac{f_{obj}}{f_{eye}} \tag{24.11}$$

The angular magnification is simply the ratio of the objective focal length to the eyepiece focal length.

Because the stars and galaxies are so distant, light-gathering power is more important to astronomers than magnification. Large light-gathering power requires a large-diameter objective lens, but large lenses are not practical; they begin to sag under their own weight. Thus **refracting telescopes,** with two lenses, are relatively small. Serious astronomy is done with a **reflecting telescope,** such as the one shown in FIGURE 24.15.

A large-diameter mirror (the *primary mirror*) focuses the rays to form a real image, but, for practical reasons, a small flat mirror (the *secondary mirror*) reflects the rays sideways before they reach a focus. This moves the primary mirror's image out to the edge of the telescope where it can be viewed by an eyepiece on the side. None of these changes affects the overall analysis of the telescope, and its angular magnification is given by Equation 24.11 if f_{obj} is replaced by f_{pri}, the focal length of the primary mirror.

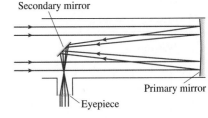
FIGURE 24.15 A reflecting telescope.

24.5 The Resolution of Optical Instruments

A camera *could* focus light with a single lens. A microscope objective *could* be built with a single lens. So why would anyone ever use a lens combination in place of a single lens? There are two primary reasons.

First, any lens has dispersion. That is, its index of refraction varies slightly with wavelength. Because the index of refraction for violet light is larger than for red light, a lens's focal length is shorter for violet light than for red light. Consequently, different colors of light come to a focus at slightly different distances from the lens. If red light is sharply focused on a viewing screen, then blue and violet wavelengths are not well focused. This imaging error, illustrated in FIGURE 24.16a, is called **chromatic aberration.**

Second, our analysis of thin lenses was based on paraxial rays traveling nearly parallel to the optical axis. A more exact analysis, taking all the rays into account, finds that rays incident on the outer edges of a spherical surface are not focused at exactly the same point as rays incident near the center. This imaging error, shown in FIGURE 24.16b, is called **spherical aberration.** Spherical aberration, which causes the image to be slightly blurred, gets worse as the lens diameter increases.

FIGURE 24.16 Chromatic aberration and spherical aberration prevent simple lenses from forming perfect images.

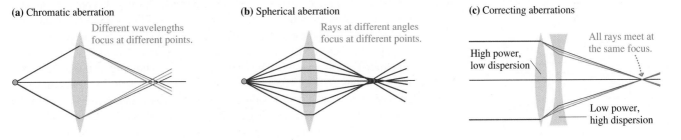

(a) Chromatic aberration
Different wavelengths focus at different points.

(b) Spherical aberration
Rays at different angles focus at different points.

(c) Correcting aberrations
High power, low dispersion
All rays meet at the same focus.
Low power, high dispersion

Fortunately, the chromatic and spherical aberrations of a converging lens and a diverging lens are in opposite directions. When a converging lens and a diverging lens are used in combination, their aberrations tend to cancel. A combination lens, such as the one in FIGURE 24.16c, can produce a much sharper focus than a single lens with the equivalent focal length. Consequently, most optical instruments use combination lenses rather than single lenses.

Diffraction Again

According to the ray model of light, a perfect lens (one with no aberrations) should be able to form a perfect image. But the ray model of light, though a very good model for lenses, is not an absolutely correct description of light. If we look closely, the wave aspects of light haven't entirely disappeared. In fact, the performance of optical equipment is limited by the diffraction of light.

FIGURE 24.17 A lens both focuses and diffracts the light passing through.

(a) A lens acts as a circular aperture.

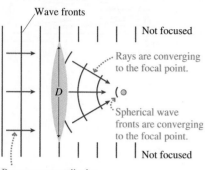

Wave fronts

Not focused

Rays are converging to the focal point.

D

Spherical wave fronts are converging to the focal point.

Not focused

Rays are perpendicular to the wave fronts.

(b) The aperture and focusing effects can be separated.

Wave fronts

Ideal diffractionless lens with focal length *f*

D

f

Circular aperture of diameter *D*

(c) The lens focuses the diffraction pattern in the focal plane.

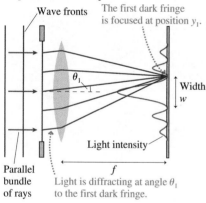

Wave fronts

The first dark fringe is focused at position y_1.

θ_1

Width *w*

Light intensity

Parallel bundle of rays

f

Light is diffracting at angle θ_1 to the first dark fringe.

FIGURE 24.17a shows a plane wave, with parallel light rays, being focused by a lens of diameter *D*. According to the ray model of light, a perfect lens would focus parallel rays to a perfect point. Notice, though, that only a piece of each wave front passes *through* the lens and gets focused. In effect, **the lens itself acts as a circular aperture** in an opaque barrier, allowing through only a portion of each wave front. Consequently, **the lens diffracts the light wave.** The diffraction is usually very small because *D* is usually much greater than the wavelength of the light; nonetheless, this small amount of diffraction is the limiting factor in how well the lens can focus the light.

FIGURE 24.17b separates the diffraction from the focusing by modeling the lens as an actual aperture of diameter *D* followed by an "ideal" diffractionless lens. You learned in Chapter 22 that a circular aperture produces a diffraction pattern with a bright central maximum surrounded by dimmer fringes. A converging lens brings this diffraction pattern to a focus in the image plane, as shown in **FIGURE 24.17c**. As a result, a perfect lens focuses parallel light rays not to a perfect point of light, as we expected, but to a small, circular diffraction pattern.

The angle to the first minimum of a circular diffraction pattern is $\theta_1 = 1.22\lambda/D$. The ray that passes through the center of a lens is not bent, so Figure 24.17c uses this ray to show that the position of the dark fringe is $y_1 = f\tan\theta_1 \approx f\theta_1$. Thus the width of the central maximum in the focal plane is

$$w_{min} \approx 2f\theta_1 = \frac{2.44\lambda f}{D} \qquad \text{(minimum spot size)} \qquad (24.12)$$

This is the **minimum spot size** to which a lens can focus light.

Lenses are often limited by aberrations, so not all lenses can focus parallel light rays to a spot this small. A well-crafted lens, for which Equation 24.12 is the minimum spot size, is called a *diffraction-limited lens*. No optical design can overcome the spreading of light due to diffraction, and it is because of this spreading that the image point has a minimum spot size. The image of an actual object, rather than of parallel rays, becomes a mosaic of overlapping diffraction patterns, so even the most perfect lens inevitably forms an image that is slightly fuzzy.

For various reasons, it is difficult to produce a diffraction-limited lens having a focal length that is much less than its diameter. The very best microscope objectives have $f \approx 0.5D$. This implies that **the smallest diameter to which you can focus a spot of light, no matter how hard you try, is $w_{min} \approx \lambda$.** This is a fundamental limit on the performance of optical equipment. Diffraction has very real consequences!

One example of these consequences is found in the manufacturing of integrated circuits. Integrated circuits are made by creating a "mask" showing all the components and their connections. A lens images this mask onto the surface of a semiconductor wafer that has been coated with a substance called *photoresist*. Bright areas in the mask expose the photoresist, and subsequent processing steps chemically etch away the exposed areas while leaving behind areas that had been in the shadows of the mask. This process is called *photolithography*.

The power of a microprocessor and the amount of memory in a memory chip depend on how small the circuit elements can be made. Diffraction dictates that a circuit element can be no smaller than the smallest spot to which light can be focused, which is roughly the wavelength of the light. If the mask is projected with ultraviolet light having $\lambda \approx 200$ nm, then the smallest elements on a chip are about 200 nm wide. This is, in fact, just about the current limit of technology.

EXAMPLE 24.7 | **Seeing stars**

A 12-cm-diameter telescope lens has a focal length of 1.0 m. What is the diameter of the image of a star in the focal plane if the lens is diffraction limited *and* if the earth's atmosphere is not a limitation?

MODEL Stars are so far away that they appear as points in space. An ideal diffractionless lens would focus their light to arbitrarily small points. Diffraction prevents this. Model the telescope lens as a 12-cm-diameter aperture in front of an ideal lens with a 1.0 m focal length.

SOLVE The minimum spot size in the focal plane of this lens is

$$w = \frac{2.44\lambda f}{D}$$

where D is the lens diameter. What is λ? Because stars emit white light, the *longest* wavelengths spread the most and determine the size of the image that is seen. If we use $\lambda = 700$ nm as the approximate upper limit of visible wavelengths, we find $w = 1.4 \times 10^{-5}$ m $= 14\ \mu$m.

ASSESS This is certainly small, and it would appear as a point to your unaided eye. Nonetheless, the spot size would be easily noticed if it were recorded on film and enlarged. Turbulence and temperature effects in the atmosphere, the causes of the "twinkling" of stars, prevent ground-based telescopes from being this good, but space-based telescopes really are diffraction limited.

Resolution

Suppose you point a telescope at two nearby stars in a galaxy far, far away. If you use the best possible detector, will you be able to distinguish separate images for the two stars, or will they blur into a single blob of light? A similar question could be asked of a microscope. Can two microscopic objects, very close together, be distinguished if sufficient magnification is used? Or is there some size limit at which their images will blur together and never be separated? These are important questions about the *resolution* of optical instruments.

Because of diffraction, the image of a distant star is not a point but a circular diffraction pattern. Our question, then, really is: How close together can two diffraction patterns be before you can no longer distinguish them? One of the major scientists of the 19th century, Lord Rayleigh, studied this problem and suggested a reasonable rule that today is called **Rayleigh's criterion.**

FIGURE 24.18 shows two distant point sources being imaged by a lens of diameter D. The angular separation between the objects, as seen from the lens, is α. Rayleigh's criterion states that

- The two objects are resolvable if $\alpha > \theta_{min}$, where $\theta_{min} = \theta_1 = 1.22\lambda/D$ is the angle of the first dark fringe in the circular diffraction pattern.
- The two objects are not resolvable if $\alpha < \theta_{min}$ because their diffraction patterns are too overlapped.
- The two objects are marginally resolvable if $\alpha = \theta_{min}$. The central maximum of one image falls exactly on top of the first dark fringe of the other image. This is the situation shown in the figure.

FIGURE 24.19 shows enlarged photographs of the images of two point sources. The images are circular diffraction patterns, not points. The two images are close but distinct where the objects are separated by $\alpha > \theta_{min}$. Two objects really were recorded in the photo at the bottom, but their separation is $\alpha < \theta_{min}$ and their images have blended together. In the middle photo, with $\alpha = \theta_{min}$, you can see that the two images are just barely resolved.

The angle

$$\theta_{min} = \frac{1.22\lambda}{D} \qquad \text{(angular resolution of a lens)} \qquad (24.13)$$

is called the **angular resolution** of a lens. The angular resolution of a telescope depends on the diameter of the objective lens (or the primary mirror) and the wavelength of the light; magnification is not a factor. Two images will remain overlapped and unresolved no matter what the magnification if their angular separation is less than θ_{min}. For visible light, where λ is pretty much fixed, the only parameter over which the astronomer has any control is the diameter of the lens or mirror of the

FIGURE 24.18 Two images that are marginally resolved.

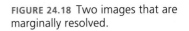
The maximum of image 2 falls on the first dark fringe of image 1. The images are marginally resolved.

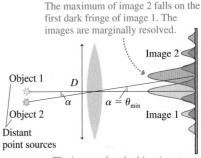

The image of each object is not a perfect point, but a small circular diffraction pattern.

FIGURE 24.19 Enlarged photographs of the images of two point sources.

$\alpha > \theta_{min}$
Resolved

$\alpha = \theta_{min}$
Marginally resolved

$\alpha < \theta_{min}$
Not resolved

The size of the features in an integrated circuit is limited by the diffraction of light.

telescope. The urge to build ever-larger telescopes is motivated, in part, by a desire to improve the angular resolution. (Another motivation is to increase the light-gathering power so as to see objects farther away.)

The performance of a microscope is also limited by the diffraction of light passing through the objective lens. Just as light cannot be focused to a spot smaller than about a wavelength, the most perfect microscope cannot resolve the features of objects that are smaller than a wavelength. Similarly, two objects separated by less than one wavelength—roughly 500 nm—will blur into a single object and cannot be resolved. Because atoms are approximately 0.1 nm in diameter, vastly smaller than the wavelength of visible or even ultraviolet light, there is no hope of ever seeing atoms with an optical microscope. This limitation is not simply a matter of needing a better design or more precise components; it is a fundamental limit set by the wave nature of the light with which we see.

STOP TO THINK 24.5 Four diffraction-limited lenses focus plane waves of light with the same wavelength λ. Rank in order, from largest to smallest, the spot sizes w_a to w_d.

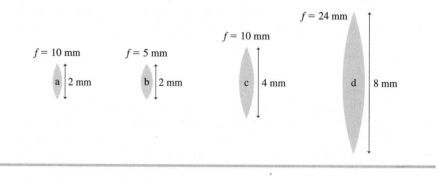

CHALLENGE EXAMPLE 24.8 | **Visual acuity**

The normal human eye has maximum visual acuity with a pupil diameter of about 3 mm. For larger pupils, acuity decreases due to increasing aberrations; for smaller pupils, acuity decreases due to increasing diffraction. If your pupil diameter is 2.0 mm, as it would be in bright light, what is the smallest-diameter circle that you should be able to see as a circle, rather than just an unresolved blob, on an eye chart at the standard distance of 20 ft? The index of refraction inside the eye is 1.33.

MODEL Assume that a 2.0-mm-diameter pupil is diffraction limited. Then the angular resolution is given by Rayleigh's criterion. Diffraction increases with wavelength, so the eye's acuity will be affected more by longer wavelengths than by shorter wavelengths. Consequently, assume that the light's wavelength in air is 600 nm.

VISUALIZE Let the diameter of the circle be d. FIGURE 24.20 shows the circle at distance $s = 20$ ft $= 6.1$ m. "Seeing the circle," shown edge-on, requires resolving the top and bottom lines as distinct.

FIGURE 24.20 Viewing a circle of diameter d.

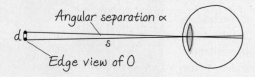

SOLVE The angular separation of the top and bottom lines of the circle is $\alpha = d/s$. Rayleigh's criterion says that a perfect lens with aperture D can just barely resolve these two lines if

$$\alpha = \frac{d}{s} = \theta_{min} = \frac{1.22\lambda_{eye}}{D} = \frac{1.22\lambda_{air}}{n_{eye}D}$$

The diffraction takes place inside the eye, where the wavelength is shortened to $\lambda_{eye} = \lambda_{air}/n_{eye}$. Thus the circle diameter that can barely be resolved with perfect vision is

$$d = \frac{1.22\lambda_{air}s}{n_{eye}D} = \frac{1.22(600 \times 10^{-9}\text{ m})(6.1\text{ m})}{(1.33)(0.0020\text{ m})} \approx 2\text{ mm}$$

That's about the height of a capital O in this book, so in principle you should—in very bright light—just barely be able to recognize it as an O at 20 feet.

ASSESS On an eye chart, the O on the line for 20/20 vision—the standard of excellent vision—is about 7 mm tall, so the calculated 2 mm, although in the right range, is a bit too small. There are three reasons. First, eye tests are done with medium-bright indoor lighting. Your acuity really does improve in light bright enough to reduce your pupil diameter to 2.0 mm. Second, although aberrations of the eye are reduced with a smaller pupil, they haven't vanished. And third, for a 2-mm-tall object at 20 ft, the size of the image on the retina is barely larger than the spacing between the cone cells, so the resolution of the "detector" is also a factor. Your eye is a very good optical instrument, but not perfect.

SUMMARY

The goal of Chapter 24 has been to understand some common optical instruments and their limitations.

Important Concepts

Lens Combinations

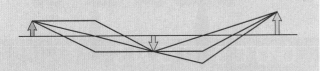

The image of the first lens acts as the object for the second lens.

Lens power: $P = \dfrac{1}{f}$ diopters, $1 \text{ D} = 1 \text{ m}^{-1}$

Resolution

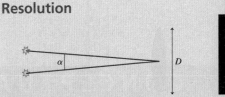

The angular resolution of a lens of diameter D is

$$\theta_{min} = 1.22\lambda/D$$

Rayleigh's criterion states that two objects separated by an angle α are marginally resolvable if $\alpha = \theta_{min}$.

Applications

Cameras

Forms a real, inverted image on a detector. The lens's **f-number** is

$$f\text{-number} = \frac{f}{D}$$

The light intensity on the detector is

$$I \propto \frac{1}{(f\text{-number})^2}$$

Magnifiers

For relaxed-eye viewing, the angular magnification is

$$M = \frac{25 \text{ cm}}{f}$$

For microscopes and telescopes, angular magnification, not lateral magnification, is the important characteristic. The eyepiece acts as a magnifier to view the image formed by the objective lens.

Vision

Refraction at the cornea is responsible for most of the focusing. The lens provides fine-tuning by changing its shape **(accommodation).**

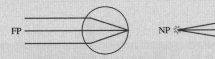

In normal vision, the eye can focus from a far point (FP) at ∞ (relaxed eye) to a near point (NP) at ≈ 25 cm (maximum accommodation).
- **Hyperopia** (farsightedness) is corrected with a converging lens.
- **Myopia** (nearsightedness) is corrected with a diverging lens.

Microscopes

The object is very close to the focal point of the objective. The total angular magnification is

$$M = -\frac{L}{f_{obj}} \frac{25 \text{ cm}}{f_{eye}}$$

The best possible spatial resolution of a microscope, limited by diffraction, is about one wavelength of light.

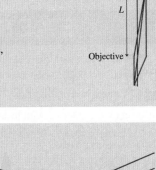

Focusing and spatial resolution

The minimum spot size to which a lens of diameter D can focus light is limited by diffraction to

$$w_{min} = \frac{2.44\lambda f}{D}$$

With the best lenses that can be manufactured, $w_{min} \approx \lambda$.

Telescopes

The object is very far from the objective.

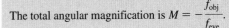

The total angular magnification is $M = -\dfrac{f_{obj}}{f_{eye}}$.

Terms and Notation

camera	iris	hyperopia	reflecting telescope
effective focal length, f	retina	myopia	chromatic aberration
aperture	accommodation	angular size	spherical aberration
f-number	far point	magnifier	minimum spot size, w_{min}
CCD	near point	angular magnification, M	Rayleigh's criterion
pixel	presbyopia	objective	angular resolution
cornea	power, P	eyepiece	
pupil	diopter, D	refracting telescope	

CONCEPTUAL QUESTIONS

1. Suppose a camera's exposure is correct when the lens has a focal length of 8.0 mm. Will the picture be overexposed, underexposed, or still correct if the focal length is "zoomed" to 16.0 mm without changing the diameter of the lens aperture? Explain.

2. A camera has a circular aperture immediately behind the lens. Reducing the aperture diameter to half its initial value will
   A. Make the image blurry.
   B. Cut off the outer half of the image and leave the inner half unchanged.
   C. Make the image less bright.
   D. All the above.
   Explain your choice.

3. Suppose you wanted special glasses designed to let you see underwater without a face mask. Should the glasses use a converging or diverging lens? Explain.

4. A friend lends you the eyepiece of his microscope to use on your own microscope. He claims the spatial resolution of your microscope will be halved, since his eyepiece has the same diameter as yours but twice the magnification. Is his claim valid? Explain.

5. A diffraction-limited lens can focus light to a 10-μm-diameter spot on a screen. Do the following actions make the spot diameter larger, make it smaller, or leave it unchanged?
   A. Decreasing the wavelength of the light.
   B. Decreasing the lens diameter.
   C. Decreasing the lens focal length.
   D. Decreasing the lens-to-screen distance.

6. To focus parallel light rays to the smallest possible spot, should you use a lens with a small f-number or a large f-number? Explain.

7. An astronomer is trying to observe two distant stars. The stars are marginally resolved when she looks at them through a filter that passes green light with a wavelength near 550 nm. Which of the following actions would improve the resolution? Assume that the resolution is not limited by the atmosphere.
   A. Changing the filter to a different wavelength. If so, should she use a shorter or a longer wavelength?
   B. Using a telescope with an objective lens of the same diameter but a different focal length. If so, should she select a shorter or a longer focal length?
   C. Using a telescope with an objective lens of the same focal length but a different diameter. If so, should she select a larger or a smaller diameter?
   D. Using an eyepiece with a different magnification. If so, should she select an eyepiece with more or less magnification?

EXERCISES AND PROBLEMS

Exercises

Section 24.1 Lenses in Combination

1. ‖ Two converging lenses with focal lengths of 40 cm and 20 cm are 10 cm apart. A 2.0-cm-tall object is 15 cm in front of the 40-cm-focal-length lens.
   a. Use ray tracing to find the position and height of the image. Do this accurately with a ruler or paper with a grid. Estimate the image distance and image height by making measurements on your diagram.
   b. Calculate the image position and height. Compare with your ray-tracing answers in part a.

2. ‖ A converging lens with a focal length of 40 cm and a diverging lens with a focal length of −40 cm are 160 cm apart. A 2.0-cm-tall object is 60 cm in front of the converging lens.
   a. Use ray tracing to find the position and height of the image. Do this accurately with a ruler or paper with a grid. Estimate the image distance and image height by making measurements on your diagram.
   b. Calculate the image position and height. Compare with your ray-tracing answers in part a.

3. ‖ A 2.0-cm-tall object is 20 cm to the left of a lens with a focal length of 10 cm. A second lens with a focal length of 15 cm is 30 cm to the right of the first lens.
   a. Use ray tracing to find the position and height of the image. Do this accurately with a ruler or paper with a grid. Estimate the image distance and image height by making measurements on your diagram.
   b. Calculate the image position and height. Compare with your ray-tracing answers in part a.

4. ‖ A 2.0-cm-tall object is 20 cm to the left of a lens with a focal length of 10 cm. A second lens with a focal length of 5 cm is 30 cm to the right of the first lens.
   a. Use ray tracing to find the position and height of the image. Do this accurately with a ruler or paper with a grid. Estimate the image distance and image height by making measurements on your diagram.
   b. Calculate the image position and height. Compare with your ray-tracing answers in part a.

5. ‖‖ A 2.0-cm-tall object is 20 cm to the left of a lens with a focal length of 10 cm. A second lens with a focal length of −5 cm is 30 cm to the right of the first lens.
   a. Use ray tracing to find the position and height of the image. Do this accurately with a ruler or paper with a grid. Estimate the image distance and image height by making measurements on your diagram.
   b. Calculate the image position and height. Compare with your ray-tracing answers in part a.

Section 24.2 The Camera

6. | A 2.0-m-tall man is 10 m in front of a camera with a 15-mm-focal-length lens. How tall is his image on the detector?

7. | What is the f-number of a lens with a 35 mm focal length and a 7.0-mm-diameter aperture?

8. | A 12-mm-focal-length lens has a 4.0-mm-diameter aperture. What is the aperture diameter of an 18-mm-focal-length lens with the same f-number?

9. | What is the aperture diameter of a 12-mm-focal-length lens set to $f/4.0$?

10. | A camera takes a properly exposed photo at $f/5.6$ and 1/125 s. What shutter speed should be used if the lens is changed to $f/4.0$?

11. ‖‖ A camera takes a properly exposed photo with a 3.0-mm-diameter aperture and a shutter speed of 1/125 s. What is the appropriate aperture diameter for a 1/500 s shutter speed?

Section 24.3 Vision

12. ‖ Ramon has contact lenses with the prescription +2.0 D.
   BIO
   a. What eye condition does Ramon have?
   b. What is his near point without the lenses?

13. | Ellen wears eyeglasses with the prescription −1.0 D.
   BIO
   a. What eye condition does Ellen have?
   b. What is her far point without the glasses?

14. | What is the f-number of a relaxed eye with the pupil fully
   BIO dilated to 8.0 mm? Model the eye as a single lens 2.4 cm in front of the retina.

Section 24.4 Optical Systems That Magnify

15. | A magnifier has a magnification of 5×. How far from the lens should an object be held so that its image is at the near-point distance of 25 cm?

16. ‖ A microscope has a 20 cm tube length. What focal-length objective will give total magnification 500× when used with a eyepiece having a focal length of 5.0 cm?

17. ‖ A standardized biological microscope has an 8.0-mm-focal-length objective. What focal-length eyepiece should be used to achieve a total magnification of 100×?

18. ‖ A 6.0-mm-diameter microscope objective has a focal length of 9.0 mm. What object distance gives a lateral magnification of −40?

19. | A 20× telescope has a 12-cm-diameter objective lens. What minimum diameter must the eyepiece lens have to collect all the light rays from an on-axis distant source?

20. ‖ A reflecting telescope is built with a 20-cm-diameter mirror having a 1.00 m focal length. It is used with a 10× eyepiece. What are (a) the magnification and (b) the f-number of the telescope?

Section 24.5 The Resolution of Optical Instruments

21. ‖ A scientist needs to focus a helium-neon laser beam ($\lambda = 633$ nm) to a 10-μm-diameter spot 8.0 cm behind a lens.
   a. What focal-length lens should she use?
   b. What minimum diameter must the lens have?

22. ‖ Two lightbulbs are 1.0 m apart. From what distance can these lightbulbs be marginally resolved by a small telescope with a 4.0-cm-diameter objective lens? Assume that the lens is diffraction limited and $\lambda = 600$ nm.

Problems

23. ‖ A 1.0-cm-tall object is located 4.0 cm to the left of a converging lens with a focal length of 5.0 cm. A diverging lens, of focal length −8.0 cm, is 12 cm to the right of the first lens. Find the position, size, and orientation of the final image.

24. | In FIGURE P24.24, are parallel rays from the left focused to a point? If so, on which side of the lens and at what distance?

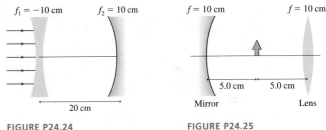

FIGURE P24.24   FIGURE P24.25

25. ‖ The rays leaving the two-component optical system of FIGURE P24.25 produce two distinct images of the 1.0-cm-tall object.
   a. What are the position (relative to the lens), orientation, and height of each image?
   b. Draw two ray diagrams, one for each image, showing how the images are formed.

26. | A common optical instrument in a laser laboratory is a *beam expander*. One type of beam expander is shown in FIGURE P24.26. The parallel rays of a laser beam of width w_1 enter from the left.
   a. For what lens spacing d does a parallel laser beam exit from the right?
   b. What is the width w_2 of the exiting laser beam?

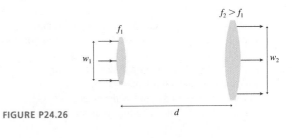

FIGURE P24.26

27. | A common optical instrument in a laser laboratory is a *beam expander*. One type of beam expander is shown in FIGURE P24.27. The parallel rays of a laser beam of width w_1 enter from the left.
    a. For what lens spacing d does a parallel laser beam exit from the right?
    b. What is the width w_2 of the exiting laser beam?

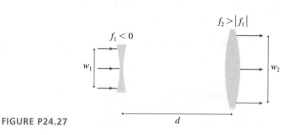

FIGURE P24.27

28. ‖ A 15-cm-focal-length converging lens is 20 cm to the right of a 7.0-cm-focal-length converging lens. A 1.0-cm-tall object is distance L to the left of the first lens.
    a. For what value of L is the final image of this two-lens system halfway between the two lenses?
    b. What are the height and orientation of the final image?
29. | A 1.0-cm-tall object is 110 cm from a screen. A diverging lens with focal length −20 cm is 20 cm in front of the object. What are the focal length and distance from the screen of a second lens that will produce a well-focused, 2.0-cm-tall image on the screen?
30. ‖ You use your 8× binoculars to focus on a 14-cm-long bird in a tree 18 m away from you. What angle (in degrees) does the image of the warbler subtend on your retina?
31. ‖ Yang can focus on objects 150 cm away with a relaxed eye.
    BIO With full accommodation, she can focus on objects 20 cm away. After her eyesight is corrected for distance vision, what will her near point be while wearing her glasses?
32. ‖ The cornea, a boundary between the air and the aqueous hu-
    BIO mor, has a 3.0 cm focal length when acting alone. What is its radius of curvature?
33. | The objective lens of a telescope is a symmetric glass lens with 100 cm radii of curvature. The eyepiece lens is also a symmetric glass lens. What are the radii of curvature of the eyepiece lens if the telescope's magnification is 20×?
34. ‖ You've been asked to build a telescope from a 2.0× magnifying lens and a 5.0× magnifying lens.
    a. What is the maximum magnification you can achieve?
    b. Which lens should be used as the objective? Explain.
    c. What will be the length of your telescope?
35. | Marooned on a desert island and with a lot of time on your hands, you decide to disassemble your glasses to make a crude telescope with which you can scan the horizon for rescuers. Luckily you're farsighted, and, like most people, your two eyes have different lens prescriptions. Your left eye uses a lens of power +4.5 D, and your right eye's lens is +3.0 D.
    a. Which lens should you use for the objective and which for the eyepiece? Explain.
    b. What will be the magnification of your telescope?
    c. How far apart should the two lenses be when you focus on distant objects?
36. ‖ You've been asked to build a 12× microscope from a 2.0× magnifying lens and a 4.0× magnifying lens.
    a. Which lens should be used as the objective?
    b. What will be the tube length of your microscope?

37. ‖ A microscope with a tube length of 180 mm achieves a total magnification of 800× with a 40× objective and a 20× eyepiece. The microscope is focused for viewing with a relaxed eye. How far is the sample from the objective lens?
38. | High-power lasers are used to cut and weld materials by focusing the laser beam to a very small spot. This is like using a magnifying lens to focus the sun's light to a small spot that can burn things. As an engineer, you have designed a laser cutting device in which the material to be cut is placed 5.0 cm behind the lens. You have selected a high-power laser with a wavelength of 1.06 μm. Your calculations indicate that the laser must be focused to a 5.0-μm-diameter spot in order to have sufficient power to make the cut. What is the minimum diameter of the lens you must install?
39. ‖‖ Once dark adapted, the pupil of your eye is approximately
    BIO 7 mm in diameter. The headlights of an oncoming car are 120 cm apart. If the lens of your eye is diffraction limited, at what distance are the two headlights marginally resolved? Assume a wavelength of 600 nm and that the index of refraction inside the eye is 1.33. (Your eye is not really good enough to resolve headlights at this distance, due both to aberrations in the lens and to the size of the receptors in your retina, but it comes reasonably close.)
40. ‖ The Hubble Space Telescope has a mirror diameter of 2.4 m. Suppose the telescope is used to photograph stars near the center of our galaxy, 30,000 light years away, using red light with a wavelength of 650 nm.
    a. What's the distance (in km) between two stars that are marginally resolved? The resolution of a reflecting telescope is calculated exactly the same as for a refracting telescope.
    b. For comparison, what is this distance as a multiple of the distance of Jupiter from the sun?
41. ‖ Alpha Centauri, the nearest star to our solar system, is 4.3 light years away. Assume that Alpha Centauri has a planet with an advanced civilization. Professor Dhg, at the planet's Astronomical Institute, wants to build a telescope with which he can find out whether any planets are orbiting our sun.
    a. What is the minimum diameter for an objective lens that will just barely resolve Jupiter and the sun? The radius of Jupiter's orbit is 780 million km. Assume $\lambda = 600$ nm.
    b. Building a telescope of the necessary size does not appear to be a major problem. What practical difficulties might prevent Professor Dhg's experiment from succeeding?

Challenge Problems

42. In FIGURE CP24.42, what are the position, height, and orientation of the final image? Give the position as a distance to the right or left of the lens.

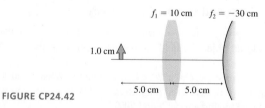

FIGURE CP24.42

43. Mars (6800 km diameter) is viewed through a telescope on a night when it is 1.1×10^8 km from the earth. Its angular size as seen through the eyepiece is 0.50°, the same size as the full moon seen by the naked eye. If the eyepiece focal length is 25 mm, how long is the telescope?

44. Your task in physics laboratory is to make a microscope from two lenses. One lens has a focal length of 2.0 cm, the other 1.0 cm. You plan to use the more powerful lens as the objective, and you want the eyepiece to be 16 cm from the objective.
    a. For viewing with a relaxed eye, how far should the sample be from the objective lens?
    b. What is the magnification of your microscope?

45. The lens shown in FIGURE CP24.45 is called an *achromatic doublet,* meaning that it has no chromatic aberration. The left side is flat, and all other surfaces have radii of curvature R.
    a. For parallel light rays coming from the left, show that the effective focal length of this two-lens system is $f = R/(2n_2 - n_1 - 1)$, where n_1 and n_2 are, respectively, the indices of refraction of the diverging and the converging lenses. Don't forget to make the thin-lens approximation.
    b. Because of dispersion, either lens alone would focus red rays and blue rays at different points. Define Δn_1 and Δn_2 as $n_{blue} - n_{red}$ for the two lenses. Find an expression for Δn_2 in terms of Δn_1 that makes $f_{blue} = f_{red}$ for the two-lens system. That is, the two-lens system does *not* exhibit chromatic aberration.
    c. Indices of refraction for two types of glass are given in the table. To make an achromatic doublet, which glass should you use for the converging lens and which for the diverging lens? Explain.

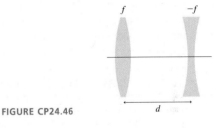

FIGURE CP24.45

	n_{blue}	n_{red}
Crown glass	1.525	1.517
Flint glass	1.632	1.616

    d. What value of R gives a focal length of 10.0 cm?

46. FIGURE CP24.46 shows a simple zoom lens in which the magnitudes of both focal lengths are f. If the spacing $d < f$, the image of the converging lens falls on the right side of the diverging lens. Our procedure of letting the image of the first lens act as the object of the second lens will continue to work in this case if we use a *negative* object distance for the second lens. This is called a *virtual object*. Consider a very distant object ($s \approx \infty$ for the first lens) and define the effective focal length as the distance from the midpoint between the lenses to the final image.
    a. Show that the effective focal length is

$$f_{eff} = \frac{f^2 - fd + \frac{1}{2}d^2}{d}$$

    b. What is the zoom for a lens that can be adjusted from $d = \frac{1}{2}f$ to $d = \frac{1}{4}f$?

FIGURE CP24.46

STOP TO THINK ANSWERS

Stop to Think 24.1: b. A diverging lens refracts rays away from the optical axis, so the rays will travel farther down the axis before converging.

Stop to Think 24.2: a. Because the shutter speed doesn't change, the *f*-number must remain unchanged. The *f*-number is f/D, so increasing f requires increasing D.

Stop to Think 24.3: a. A magnifier is a converging lens. Converging lenses are used to correct hyperopia.

Stop to Think 24.4: b. If the objective magnification is halved, the eyepiece magnification must be doubled. $M_{eye} = 25 \text{ cm}/f_{eye}$, so doubling M_{eye} requires halving f_{eye}.

Stop to Think 24.5: $w_a > w_d > w_b = w_c$. The spot size is proportional to f/D.

SUMMARY
Waves and Optics

We end our study of waves a long distance from where we started. Who would have guessed, as we examined our first pulse on a string, that we would end up discussing the resolution of microscopes? But despite the wide disparity between string waves, sound waves, and light waves, a few key ideas have stayed with us throughout Part V: the principle of superposition, interference and diffraction, and standing waves. As part of your final study of waves, you should trace the influence of these ideas through the chapters of Part V.

One point we have tried to emphasize is the *unity* of wave physics. We did not need separate theories of string waves and sound waves and light waves. Instead, a few basic ideas enabled us to understand waves of all types. By focusing on similarities, we have been able to analyze vibrating guitar strings and antireflection coatings on lenses in a single part of this book.

Unfortunately, the physics of waves is not as easily summarized as the physics of particles. Newton's laws and the conservation laws are two very general sets of principles about particles, principles that allowed us to develop the powerful problem-solving strategies of Parts I and II. You probably noticed that we have not found any general problem-solving strategies for wave problems.

This is not to say that wave physics has no structure. Rather, the knowledge structure of waves and optics rests more heavily on *phenomena* than on general principles. Unlike the knowledge structure of Newtonian mechanics, which was a "pyramid of ideas," the knowledge structure of waves is a logical grouping of the major topics you studied. This is a different way of structuring knowledge, but it still provides you with a mental framework for analyzing and thinking about wave problems.

KNOWLEDGE STRUCTURE V **Waves and Optics**

ESSENTIAL CONCEPTS	Wave speed, wavelength, frequency, phase, wave front, and ray
BASIC GOALS	What are the distinguishing features of waves?
	How does a wave travel through a medium?
	How does a medium respond to the presence of more than one wave?
	What is light and what are its properties?
GENERAL PRINCIPLES	Principle of superposition
	$v = \lambda f$ for periodic waves

Traveling Waves

- The wave speed v is a property of the medium.
- The motion of particles in the medium is distinct from the motion of the wave.
- Snapshot graphs and history graphs show the same wave from different perspectives.
- The Doppler effect of shifted frequencies is observed whenever the wave source or the detector is moving.

Standing Waves

- Standing waves are the superposition of waves moving in opposite directions.
- Nodes and antinodes are spaced by $\lambda/2$.
- Only certain discrete frequencies are allowed, depending on the boundary conditions.

Antinodes

Nodes

Interference

- Interference is constructive—crests align with crests—if two waves are in phase: $\Delta\phi = 0, 2\pi, 4\pi, \ldots$. The wave is enhanced.
- Interference is destructive—crests align with troughs—if two waves are out of phase: $\Delta\phi = \pi, 3\pi, 5\pi, \ldots$. The wave is reduced.
- The phase difference depends on the path-length difference Δr and on any phase difference between the sources.
- Beats occur when $f_1 \neq f_2$.

Light and Optics

- The wave model, used for interference and diffraction, is appropriate when apertures are comparable in size to the wavelength.

Single-slit diffraction:

Double-slit interference:

- The ray model, used for mirrors and lenses, is appropriate when apertures are much larger than the wavelength.
- Diffraction, a wave effect, limits the best possible resolution of a lens.

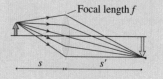

Focal length f

s s'

Tsunami!

In December 2004, an earthquake off the Indonesian coast produced a devastating water wave, a *tsunami,* that caused tremendous destruction and loss of life around the edges of the Indian Ocean, often thousands of miles from the earthquake's epicenter. The tsunami was a dramatic reminder of the power of the earth's forces and an impressive illustration of the energy carried by waves.

The Indian Ocean tsunami of 2004 was caused when a very large earthquake disrupted the seafloor along a fault line, pushing one side of the fault up several meters. This dramatic shift in the seafloor produced an almost instantaneous rise in the surface of the ocean above, much like giving a quick shake to one end of a rope. This was the disturbance that produced the tsunami. And just as shaking one end of rope causes a pulse to travel along it, the resulting water wave propagated throughout the Indian Ocean, as we see in the figure, carrying energy from the earthquake.

This computer simulation of the tsunami looks much like the ripples that spread out when you drop a pebble into a pond, but on an immensely larger scale. The individual wave pulses are up to 100 km wide, and the leading wave front spans more than 5000 km.

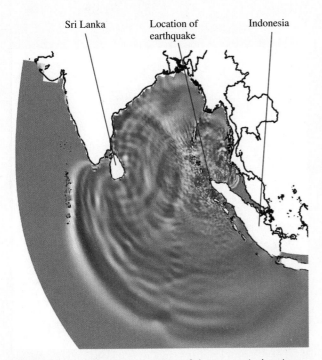

Sri Lanka   Location of earthquake   Indonesia

A frame from a computer simulation of the tsunami, showing the Indian Ocean about three hours after the earthquake. Notice the interference pattern to the east of Sri Lanka, where incoming waves and reflected waves are superimposed.

Technically, a tsunami is a "shallow-water wave," even in the deep ocean, because the scale of the wave (roughly 100 km) is much larger than the depth of the ocean (typically 4 km). Consequently, a tsunami travels differently than normal ocean waves. Unlike normal waves on the surface, whose speed is independent of depth, the speed of a shallow-water wave is determined by the depth of the ocean: The greater the depth, the greater the speed. In the deep ocean, a tsunami travels at hundreds of kilometers per hour, about the speed of a jet plane. This great speed allows a tsunami to cross oceans in only a few hours.

The height of the tsunami as it raced across the open ocean was about half a meter. Why should such a small wave—one that ships didn't even notice as it passed—be so fearsome? It's the width of the wave that matters. The wave pulse may have been only half a meter high, but it was about 100 km wide. In other words, the tsunami far from land was a half-meter-high, 100-km-wide wall of water. This is a tremendous amount of water displaced upward, and thus the tsunami was carrying a tremendous amount of energy.

As a tsunami nears shore, the ocean depth decreases and—because its speed is determined by depth—the tsunami begins to slow. This is when the awesome power of a tsunami begins to become apparent. As the leading edge of the wave slows, the trailing edge, still 100 km away and traveling much faster in deeper water, quickly begins to catch up. Water is nearly incompressible. As the width of the wave pulse decreases, the water begins to pile up higher and higher and the wave increases dramatically in height. The Indian Ocean tsunami had a height of up to 15 m (50 ft) as it came ashore.

Despite its height, a tsunami doesn't break and crash on the beach like a normal wave. The wave pulse may have narrowed dramatically from its 100 km width in the open ocean, but it is still several kilometers wide. Thus a tsunami reaching shore is more like a huge water surge than a typical wave—a wall of water that moves onto the shore and just keeps on coming. In many places, the Indian Ocean tsunami reached 2 km inland.

The impact of the Indian Ocean tsunami was devastating, but it was the first tsunami for which scientists were able to use satellites and ocean sensors to make planet-wide measurements. An analysis of the data, including computer simulations like the one seen here, has helped us better understand the physics of these ocean waves. We won't be able to stop future tsunamis, but with a better knowledge of how they are formed and how they travel, we will be better able to warn people to get out of their way.

Mathematics Review

Algebra

Using exponents:

$$a^{-x} = \frac{1}{a^x} \qquad a^x a^y = a^{(x+y)} \qquad \frac{a^x}{a^y} = a^{(x-y)} \qquad (a^x)^y = a^{xy}$$

$$a^0 = 1 \qquad a^1 = a \qquad a^{1/n} = \sqrt[n]{a}$$

Fractions:

$$\left(\frac{a}{b}\right)\left(\frac{c}{d}\right) = \frac{ac}{bd} \qquad \frac{a/b}{c/d} = \frac{ad}{bc} \qquad \frac{1}{1/a} = a$$

Logarithms:

If $a = e^x$, then $\ln(a) = x \qquad \ln(e^x) = x \qquad e^{\ln(x)} = x$

$$\ln(ab) = \ln(a) + \ln(b) \qquad \ln\left(\frac{a}{b}\right) = \ln(a) - \ln(b) \qquad \ln(a^n) = n\ln(a)$$

The expression $\ln(a + b)$ cannot be simplified.

Linear equations: The graph of the equation $y = ax + b$ is a straight line. a is the slope of the graph. b is the y-intercept.

Proportionality: To say that y is proportional to x, written $y \propto x$, means that $y = ax$, where a is a constant. Proportionality is a special case of linearity. A graph of a proportional relationship is a straight line that passes through the origin. If $y \propto x$, then

$$\frac{y_1}{y_2} = \frac{x_1}{x_2}$$

Slope $a = \dfrac{\text{rise}}{\text{run}} = \dfrac{\Delta y}{\Delta x}$

y-intercept $= b$

Quadratic equation: The quadratic equation $ax^2 + bx + c = 0$ has the two solutions $x = \dfrac{-b \pm \sqrt{b^2 - 4ac}}{2a}$.

Geometry and Trigonometry

Area and volume:

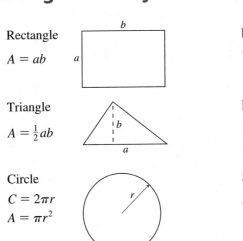

Rectangle
$$A = ab$$

Triangle
$$A = \tfrac{1}{2}ab$$

Circle
$$C = 2\pi r$$
$$A = \pi r^2$$

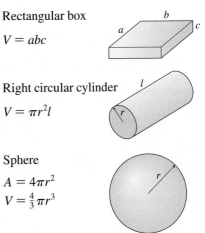

Rectangular box
$$V = abc$$

Right circular cylinder
$$V = \pi r^2 l$$

Sphere
$$A = 4\pi r^2$$
$$V = \tfrac{4}{3}\pi r^3$$

Arc length and angle: The angle θ in radians is defined as $\theta = s/r$.

The arc length that spans angle θ is $s = r\theta$.

2π rad $= 360°$

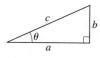

Right triangle: Pythagorean theorem $c = \sqrt{a^2 + b^2}$ or $a^2 + b^2 = c^2$

$$\sin\theta = \frac{b}{c} = \frac{\text{far side}}{\text{hypotenuse}} \qquad \theta = \sin^{-1}\left(\frac{b}{c}\right)$$

$$\cos\theta = \frac{a}{c} = \frac{\text{adjacent side}}{\text{hypotenuse}} \qquad \theta = \cos^{-1}\left(\frac{a}{c}\right)$$

$$\tan\theta = \frac{b}{a} = \frac{\text{far side}}{\text{adjacent side}} \qquad \theta = \tan^{-1}\left(\frac{b}{a}\right)$$

General triangle: $\alpha + \beta + \gamma = 180° = \pi$ rad

Law of cosines $c^2 = a^2 + b^2 - 2ab\cos\gamma$

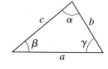

Identities:

$$\tan\alpha = \frac{\sin\alpha}{\cos\alpha} \qquad\qquad \sin^2\alpha + \cos^2\alpha = 1$$

$$\sin(-\alpha) = -\sin\alpha \qquad\qquad \cos(-\alpha) = \cos\alpha$$

$$\sin(\alpha \pm \beta) = \sin\alpha\cos\beta \pm \cos\alpha\sin\beta \qquad \cos(\alpha \pm \beta) = \cos\alpha\cos\beta \mp \sin\alpha\sin\beta$$

$$\sin(2\alpha) = 2\sin\alpha\cos\alpha \qquad\qquad \cos(2\alpha) = \cos^2\alpha - \sin^2\alpha$$

$$\sin(\alpha \pm \pi/2) = \pm\cos\alpha \qquad\qquad \cos(\alpha \pm \pi/2) = \mp\sin\alpha$$

$$\sin(\alpha \pm \pi) = -\sin\alpha \qquad\qquad \cos(\alpha \pm \pi) = -\cos\alpha$$

Expansions and Approximations

Binomial expansion: $$(1 + x)^n = 1 + nx + \frac{n(n-1)}{2}x^2 + \cdots$$

Binomial approximation: $$(1 + x)^n \approx 1 + nx \quad \text{if} \quad x \ll 1$$

Trigonometric expansions: $$\sin\alpha = \alpha - \frac{\alpha^3}{3!} + \frac{\alpha^5}{5!} - \frac{\alpha^7}{7!} + \cdots \quad \text{for } \alpha \text{ in rad}$$

$$\cos\alpha = 1 - \frac{\alpha^2}{2!} + \frac{\alpha^4}{4!} - \frac{\alpha^6}{6!} + \cdots \quad \text{for } \alpha \text{ in rad}$$

Small-angle approximation: If $\alpha \ll 1$ rad, then $\sin\alpha \approx \tan\alpha \approx \alpha$ and $\cos\alpha \approx 1$.

The small-angle approximation is excellent for $\alpha < 5°$ (≈ 0.1 rad) and generally acceptable up to $\alpha \approx 10°$.

Calculus

The letters a and n represent constants in the following derivatives and integrals.

Derivatives

$$\frac{d}{dx}(a) = 0$$

$$\frac{d}{dx}(ax) = a$$

$$\frac{d}{dx}\left(\frac{a}{x}\right) = -\frac{a}{x^2}$$

$$\frac{d}{dx}(ax^n) = anx^{n-1}$$

$$\frac{d}{dx}\left(\ln(ax)\right) = \frac{1}{x}$$

$$\frac{d}{dx}(e^{ax}) = ae^{ax}$$

$$\frac{d}{dx}\left(\sin(ax)\right) = a\cos(ax)$$

$$\frac{d}{dx}\left(\cos(ax)\right) = -a\sin(ax)$$

Integrals

$$\int x\,dx = \frac{1}{2}x^2$$

$$\int x^2\,dx = \frac{1}{3}x^3$$

$$\int \frac{1}{x^2}\,dx = -\frac{1}{x}$$

$$\int x^n\,dx = \frac{x^{n+1}}{n+1} \qquad n \neq -1$$

$$\int \frac{dx}{x} = \ln x$$

$$\int \frac{dx}{a+x} = \ln(a+x)$$

$$\int \frac{x\,dx}{a+x} = x - a\ln(a+x)$$

$$\int \frac{dx}{\sqrt{x^2 \pm a^2}} = \ln\left(x + \sqrt{x^2 \pm a^2}\right)$$

$$\int \frac{x\,dx}{\sqrt{x^2 \pm a^2}} = \sqrt{x^2 \pm a^2}$$

$$\int \frac{dx}{x^2 + a^2} = \frac{1}{a}\tan^{-1}\left(\frac{x}{a}\right)$$

$$\int \frac{dx}{(x^2 + a^2)^2} = \frac{1}{2a^3}\tan^{-1}\left(\frac{x}{a}\right) + \frac{x}{2a^2(x^2 + a^2)}$$

$$\int \frac{dx}{(x^2 \pm a^2)^{3/2}} = \frac{\pm x}{a^2\sqrt{x^2 \pm a^2}}$$

$$\int \frac{x\,dx}{(x^2 \pm a^2)^{3/2}} = -\frac{1}{\sqrt{x^2 \pm a^2}}$$

$$\int e^{ax}\,dx = \frac{1}{a}e^{ax}$$

$$\int xe^{-x}\,dx = -(x+1)e^{-x}$$

$$\int x^2 e^{-x}\,dx = -(x^2 + 2x + 2)e^{-x}$$

$$\int \sin(ax)\,dx = -\frac{1}{a}\cos(ax)$$

$$\int \cos(ax)\,dx = \frac{1}{a}\sin(ax)$$

$$\int \sin^2(ax)\,dx = \frac{x}{2} - \frac{\sin(2ax)}{4a}$$

$$\int \cos^2(ax)\,dx = \frac{x}{2} + \frac{\sin(2ax)}{4a}$$

$$\int_0^\infty x^n e^{-ax}\,dx = \frac{n!}{a^{n+1}}$$

$$\int_0^\infty e^{-ax^2}\,dx = \frac{1}{2}\sqrt{\frac{\pi}{a}}$$

Periodic Table of Elements

Key:

27	← Atomic number
Co	← Symbol
58.9	← Atomic mass

Period	1	2	3	4	5	6	7	8	9	10	11	12	13	14	15	16	17	18
1	1 H 1.0																	2 He 4.0
2	3 Li 6.9	4 Be 9.0											5 B 10.8	6 C 12.0	7 N 14.0	8 O 16.0	9 F 19.0	10 Ne 20.2
3	11 Na 23.0	12 Mg 24.3											13 Al 27.0	14 Si 28.1	15 P 31.0	16 S 32.1	17 Cl 35.5	18 Ar 39.9
4	19 K 39.1	20 Ca 40.1	21 Sc 45.0	22 Ti 47.9	23 V 50.9	24 Cr 52.0	25 Mn 54.9	26 Fe 55.8	27 Co 58.9	28 Ni 58.7	29 Cu 63.5	30 Zn 65.4	31 Ga 69.7	32 Ge 72.6	33 As 74.9	34 Se 79.0	35 Br 79.9	36 Kr 83.8
5	37 Rb 85.5	38 Sr 87.6	39 Y 88.9	40 Zr 91.2	41 Nb 92.9	42 Mo 95.9	43 Tc [98]	44 Ru 101.1	45 Rh 102.9	46 Pd 106.4	47 Ag 107.9	48 Cd 112.4	49 In 114.8	50 Sn 118.7	51 Sb 121.8	52 Te 127.6	53 I 126.9	54 Xe 131.3
6	55 Cs 132.9	56 Ba 137.3	71 Lu 175.0	72 Hf 178.5	73 Ta 180.9	74 W 183.9	75 Re 186.2	76 Os 190.2	77 Ir 192.2	78 Pt 195.1	79 Au 197.0	80 Hg 200.6	81 Tl 204.4	82 Pb 207.2	83 Bi 209.0	84 Po [209]	85 At [210]	86 Rn [222]
7	87 Fr [223]	88 Ra [226]	103 Lr [262]	104 Rf [265]	105 Db [268]	106 Sg [271]	107 Bh [272]	108 Hs [270]	109 Mt [276]	110 Ds [281]	111 Rg [280]	112 Cn [285]	113	114	115	116	117	118

Transition elements (Groups 3–12)

Inner transition elements

Lanthanides 6	57 La 138.9	58 Ce 140.1	59 Pr 140.9	60 Nd 144.2	61 Pm 144.9	62 Sm 150.4	63 Eu 152.0	64 Gd 157.3	65 Tb 158.9	66 Dy 162.5	67 Ho 164.9	68 Er 167.3	69 Tm 168.9	70 Yb 173.0
Actinides 7	89 Ac [227]	90 Th 232.0	91 Pa 231.0	92 U 238.0	93 Np [237]	94 Pu [244]	95 Am [243]	96 Cm [247]	97 Bk [247]	98 Cf [251]	99 Es [252]	100 Fm [257]	101 Md [258]	102 No [259]

An atomic mass in brackets is that of the longest-lived isotope of an element with no stable isotopes.

ActivPhysics OnLine™ Activities (MP)® www.masteringphysics.com

The following list gives the activity numbers and titles of the ActivPhysics activities available in the Pearson eText and the Study Area of MasteringPhysics, followed by the corresponding textbook page references.

APPENDIX

C

A-5

PhET Simulations (MP) www.masteringphysics.com

The following list gives the titles of the PhET simulations available in the Pearson eText and in the Study Area of MasteringPhysics, with the corresponding textbook section and page references.

*Indicates an associated tutorial is available in the MasteringPhysics Item Library.

Answers

Answers to Odd-Numbered Exercises and Problems

Chapter 20

1. 110 N
3. 2.0 m
5.
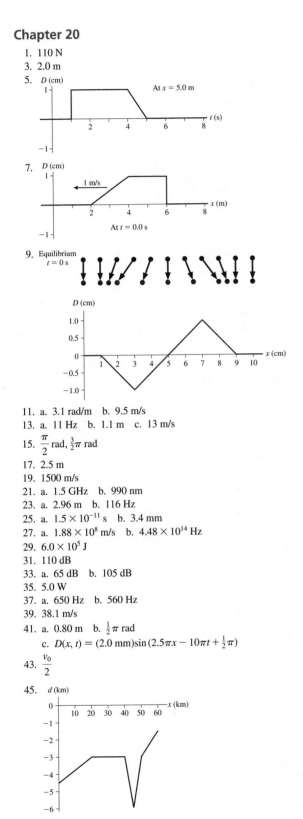
11. a. 3.1 rad/m   b. 9.5 m/s
13. a. 11 Hz   b. 1.1 m   c. 13 m/s
15. $\frac{\pi}{2}$ rad, $\frac{3}{2}\pi$ rad
17. 2.5 m
19. 1500 m/s
21. a. 1.5 GHz   b. 990 nm
23. a. 2.96 m   b. 116 Hz
25. a. 1.5×10^{-11} s   b. 3.4 mm
27. a. 1.88×10^8 m/s   b. 4.48×10^{14} Hz
29. 6.0×10^5 J
31. 110 dB
33. a. 65 dB   b. 105 dB
35. 5.0 W
37. a. 650 Hz   b. 560 Hz
39. 38.1 m/s
41. a. 0.80 m   b. $\frac{1}{2}\pi$ rad
    c. $D(x, t) = (2.0 \text{ mm})\sin(2.5\pi x - 10\pi t + \frac{1}{2}\pi)$
43. $\dfrac{v_0}{2}$
45.

47. 410 ms
49. a. 440 Hz   b. 3.4 m
51. a. $-y$-direction   b. y-axis   c. 0.701 m, 350 m/s, 2.00 ms
53. a. 12.6 N   b. 2.00 cm   c. 12.8 m/s
55. $D(x, t) = (0.010 \text{ mm})\sin[(\pi \text{ rad/m})x - (400\pi \text{ rad/s})t + \frac{1}{2}\pi \text{ rad}]$
57. -19 m/s, 0 m/s, 19 m/s
59. 8
61. 9.4 m/s
63. a. 0.095 W/m^2 b. 1.6 MW/m^2
65. a. 6.67×10^4 W   b. 8.5×10^{10} W/m^2
67. 50 m
69. 1.3
71. 21 min
75. Receding at 1.5×10^6 m/s
77. 0.07°C
81. 29 s

Chapter 21

1.
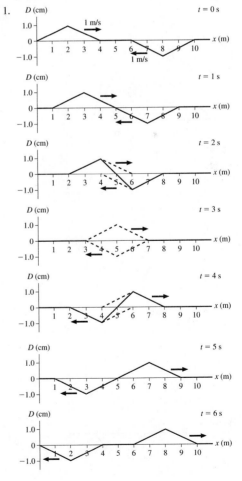

3. D (cm)

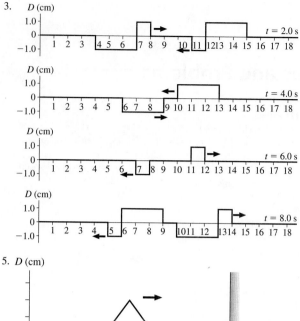

5. D (cm)

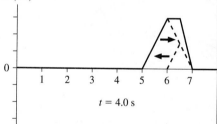

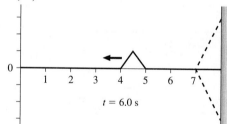

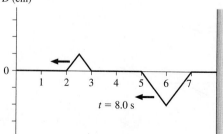

7. 50 Hz
9. a. 4.8 m, 2.4 m, 1.6 m   b. 75 Hz
11. 12 kg
13. a. 2.42 m, 1.21 m, 0.807 m   b. 4.84 m, 1.61 m, 0.968 m
15. 512 Hz
17. 2180 N
19. a. 80 cm b. 100 cm
21. 216 nm
23. a. In phase

b.

	r_1	r_2	Δr	C/D
P	3λ	4λ	λ	C
Q	$\frac{7}{2}\lambda$	2λ	$\frac{3}{2}\lambda$	D
R	$\frac{5}{2}\lambda$	$\frac{7}{2}\lambda$	λ	C

25. Perfect destructive
27. 203 Hz
29. 1.26 cm
31. $A(x = 10 \text{ cm}) = 0.62$ cm, $A(x = 20 \text{ cm}) = 1.18$ cm,
    $A(x = 30 \text{ cm}) = 1.62$ cm, $A(x = 40 \text{ cm}) = 1.90$ cm,
    $A(x = 50 \text{ cm}) = 2.00$ cm
33. 1.4 cm
35. 180 Hz
37. 28.4 cm
39. 18 cm
41. 140 N/m
43. 6.1 cm
45. $9\mu_0/4$
47. 13.0 cm
49. 580 Hz, 4.9 kHz
51. 12.1 kHz
53. 450 N
55. 93 m
57. 7.9 cm
59. a. 850 Hz   b. $-\pi/2$ rad
61. 7.2 cm
63. 20
65. 170 Hz
67. 1/3
69. a. a b. 1.0 m   c. 9
71. a. 5   b. 4.6 mm
73. 7.0 m/s
75. 4.0 cm, 35 cm, 65 cm
77. 2.0 kg
79. a. $\lambda_1 = 20.0$ m, $\lambda_2 = 10.0$ m, $\lambda_3 = 6.67$ m
    b. $v_1 = 5.59$ m/s, $v_2 = 3.95$ m/s, $v_3 = 3.22$ m/s
    d. $T_1 = 3.58$ s, $T_2 = 2.53$ s, $T_3 = 2.07$ s

Chapter 22

1. 0.023 rad = 1.3°
3. 1000 nm
5. 0.36 mm
7. 0.286°
9. 1.6°, 3.2°
11. 530
13. 7.9 μm
15. 0.20 mm
17. 0.50 mm
19. 4.0 mm
21. 7.6 m
23. 0.015 rad = 0.87°
25. 0.25 mm
27. 400 nm

29. 0.2895 mm
31. a. Single slit   b. 0.15 mm
33. 1.67 m
35. 3 mW/m^2
37. 12.0 μm
39. 667.8 nm
41. 25 cm
43. 3
45. a. 1230 lines/mm   b. 46.5°
47. 670 lines/mm
49. 16°
51. 800 lines/mm
53. a. 2   b. 1.15   c. 1
55. 670 nm
57. 0.12 mm
59. a. 550 nm   b. 0.40 mm
61. 50 cm
63. a. 22.3°   b. 16.6°
65. 19
67. a. Dark   b. 1.597
69. a. No   b. 0.044°   c. 4.6 mm   d. 1.5 m
71. b. 0.022°, 0.058°
73. b. −11.5°, −53.1°
75. a. 0.52 mm   b. 0.074°   c. 1.3 m

Chapter 23

1. a. 3.3 ns   b. 75 cm, 67 cm, 46 cm
3. 0.40 ns
5. 30°
7. 6.1 m
9. 433 cm
11. 16°
13. 1.39
15. 76.7°
17. 3.2 cm
19. 1.52
21. 1.48
23. 1600 nm
25. 6.0 cm behind the lens, inverted
27. 7.5 cm in front of the lens, upright
29. 68 cm
31. 200 cm
33. 36 cm
35. 40 cm in front of mirror, inverted
37. 12 cm behind mirror, upright
39. a. 3   b. B(+1.0 m, −2.0 m), C(−1.0 m, +2.0 m), D(+1.0 m, +2.0 m)
41. 10 m
43. 1.7

45. a. 87 cm   b. 65 cm   c. 43 cm
47. 4.0 m
49. a. Total internal reflection   b. Refraction at 72°   c. 18 cm
51. 1.58
53. 1.0°
55. 2.00
57. b. −15 cm, 1.5 cm, agree
59. b. 50 cm, 0.67 cm, agree
61. b. −20 cm, 0.33 cm, agree
63. 15.1 cm
65. −15 cm, 0.75 cm, behind, upright
67. Concave, 3.6 cm
69. 67 cm, 1.0 m
71. a. 5.9 cm   b. 6.0 cm
73. 16 cm
75. 13 cm
79. a. $t = \dfrac{n_1}{c}\sqrt{x^2 + a^2} + \dfrac{n_2}{c}\sqrt{(w-x)^2 + b^2}$

   b. $0 = \dfrac{n_1 x}{c\sqrt{x^2 + a^2}} - \dfrac{n_2(w-x)}{c\sqrt{(w-x)^2 + b^2}}$

81. b. 1.574

Chapter 24

1. b. $s'_2 = 49$ cm, $h'_2 = 4.6$ cm
3. b. $s'_2 = 30$ cm, $h'_2 = 6.0$ cm
5. b. $s'_2 = -3.33$ cm, $h'_2 = 0.66$ cm
7. 5.0
9. 3.0 mm
11. 6.0 mm
13. a. Myopia   b. 100 cm
15. 6.3 cm
17. 5.0 cm
19. 6.0 mm
21. a. 8.0 cm   b. 1.2 cm
23. Upright image, 1.0 cm tall, 6.4 cm to left of the second lens
25. a. Both images 2.0 cm tall; one upright 10 cm left of lens, the other inverted 20 cm to right of lens.
27. a. $f_2 + f_1$ b. $\dfrac{f_2}{|f_1|}w_1$
29. 16 cm placed 80 cm from screen
31. 23 cm
33. 5.0 cm
35. a. +3.0 D as objective   b. −1.5   c. 0.56 m
37. 4.6 mm
39. 15 km
41. a. 3.8 cm   b. Sun is too bright
43. 3.5 m
45. b. $\Delta n_2 = \dfrac{1}{2}\Delta n_1$ c. Crown converging, flint diverging   d. 4.18 cm

Credits

Index